To Roy. Hoping you find this an
encouragement and informative book,
in your future exploits
of the Simple Life.

From Ross

AF574152

Fish Farming

Garden Farming Series

Fish Farming

Cyril C. Harris

PELHAM BOOKS

First published in Great Britain by
Pelham Books Ltd
52 Bedford Square
London WC1B 3EF
1978

© 1978 Cyril C. Harris

All Rights Reserved. No part of this publication may be reproduced, stored in a retrieval system, or transmitted in any form or by any means, electronic, mechanical, photocopying or otherwise, without the prior permission of the Copyright owner.

ISBN 0 7207 1040 5

Filmset and printed in Great Britain by
BAS Printers Limited, Over Wallop, Hampshire
and bound by Redwood Burn, Esher

Contents

Acknowledgements

I would like to acknowledge the great assistance of Mr Keith Sangster, Joint Managing Director of Thompson and Morgan (Ipswich) Ltd, who has pioneered the idea of garden fish farming and was responsible for the introduction of the low energy aerator; Mr R. G. A. Davies of Stapeley Water Gardens Ltd, for his valued advice and guidance with illustrations; Abu (Great Britain) Ltd, who supplied a smoker for trials, together with detailed literature; *The Field* and *Trout and Salmon* for valuable assistance in compiling the list of commercial fish farms.

Picture Acknowledgements
Thompson and Morgan (Ipswich) Ltd—p. 44;
Stapeley Water Gardens Ltd—pp. 47, 48, 50, 53, 79;
Abu (Great Britain) Ltd—p. 101.

Introduction

In these days of fast-increasing inflation and wage restriction, families are finding it essential to do everything they can to reduce their cost of living. Growing one's own fruit and vegetables has become necessary, as well as pleasurable, for many of us. Home freezers are now very popular, and, if used intelligently, do not take long to repay their initial cost. You can grow your own produce cheaply from seeds or plants, pick it at its peak, eat what you want, and freeze the rest so that you can still enjoy it in its prime when it is out of season, and without having the trouble of going to the shops.

But a far larger part of the average family's food bill goes on protein foods, which, like fruit and vegetables, are an important part of a healthy, well-balanced diet. Cheese and eggs are still relatively cheap sources of protein, but most of us also like to eat meat and fish, whose prices have risen phenomenally. Even the 'cheaper' cuts of meat and types of fish are now expensive, and a home-produced source of these foods is the best way to make the maximum savings.

Traditionally, these sources were chickens, which produce eggs, of course, as well as meat, and rabbits. But nowadays people are also turning to fresh-water fish such as rainbow trout, carp and tench, which can be raised in ponds and ornamental garden pools. These fish make attractive substitutes for the unproductive goldfish, golden orfe, golden rudd, etc.

Although to many people this may sound a novel idea, it is nothing of the sort. All the medieval monasteries had within their walls one or more fishponds stocked with carp. The monks were obliged to eat fish on certain days for religious reasons, and sea fish would have been inedible after a long journey from the coast. A fish pool was also a feature of many of the gardens of great houses, and, before the eighteenth century, served the same purpose. Then the style of large houses and their gardens changed—the pool became ornamental and was planted out with water lilies, and goldfish were introduced. In the more rural districts of France many small pools are stocked with carp still, and provide fish for the table.

Before discussing garden fish farming in greater detail, let us consider the alternative methods of producing protein foods at home, and their advantages and disadvantages.

Keeping chickens in the back yard has been common practice for a very long time. Although to some extent they are kept for their meat, this is usually only secondary to their ability to provide eggs, which are still fairly cheap as sources of protein. Are chickens, then, really worth keeping? They take up quite a lot of space in the garden. Chicken sheds and runs are not particularly attractive to look at. The birds have to be fed regularly, at least twice a day, including holiday times. Their food must either be prepared by you, or purchased—which is not cheap. Their housing must be kept dry, warm and clean. Your neighbours may not like the noise or the smell.

Rabbits are simpler to feed, silent, and more cost-effective as sources of meat. But apart from that all the disadvantages of keeping chickens apply to them, too.

Now consider the advantages of raising fish in your garden. Sea fish, and even fresh-water fish such as rainbow trout, can never be as fresh on the fishmonger's slab as when they are newly caught. The balance and beauty of your garden need not be disturbed. An existing ornamental pool, providing it has certain minimal dimensions, can be used. And you can still enjoy the sight of water lilies and other aquatic plants, because fresh-water edible fish are no more harmful to them than goldfish. If you don't already have a pool, they are inexpensive and fairly easy to build nowadays. Some people are fortunate enough to have an existing natural pool, which would make an excellent home for fresh-water fish.

Needless to say, keeping fish does not cause any nuisance to the neighbours. There is no objectionable smell, and fish themselves make no noise. The maximum noise would be the very small amount given out by a $\frac{1}{4}$-hp electric motor driving a water pump or an aerator. There is no daily cleaning out to do. All the excreta is retained in the water, and so does not smell. In fact, it seals the surface of the pool after a time and helps to make it waterproof. The excreta content makes fish-pool water very nutritious for garden plants, and it is therefore an excellent liquid manure.

The fish can be fed with table scraps. If ornamental water plants are grown, they will provide a certain amount of food. For carnivorous fish, the average garden pool will also provide an adequate supply of insects. Occasionally food tablets may be required—if other sources of food dry up, for instance when you return from holiday. The average-length family holiday presents no problems as far as fish are concerned, because most pools provide enough vegetable and animal matter to keep the fish happy for some time.

Fish in a garden pool also prevent midges and mosquitoes from breeding. They will kill aphids and, very important in an ornamental pool, they will eliminate water-lily beetles which damage the leaves of these plants. This works particularly well if the pool has one or more strong jets of water—a fountain, for instance, playing on the surface of the water. The force of the water will dislodge the adult beetles and larvae from the foliage and the fish will eat them. From an aesthetic point of view, water and fish are undoubtedly of great value in any garden. They provide interest, movement and reflection.

In 1952 Dr Magnus Pyke, the well-known food scientist and broadcaster, wrote this in a book called *Townsman's Food*:

> Fish is an anomaly in the modern world by being the only remaining wild thing used by civilized man as a major food-stuff. All other foods, animal and vegetable alike, have been domesticated. Fish alone fends for itself without benefit of farm pasture, fertilizer or (save for a few specialized trout hatcheries) purposeful breeding. It is, therefore, worth considering its food supply and biological origin, and it is also of interest to dwell for a moment on the theoretical possibilities which might arise if this anachronistic savage were, like all others, put to work for civilized mankind. Many unhappy precedents suggest that, if this does not take place, the alternative may be extinction.

Generally speaking, the position has not changed very much since this was written. Both sea fish and fresh-water fish still run wild. Man uses very sophisticated methods to catch them, even to the extent of fishing out the abundant sources, while he does nothing to nourish and foster fish, except to apply restrictions to certain areas in an attempt to reduce the decline of the fish population temporarily. Even then, these precautions are more often taken in the national interest than in that of mankind generally. Admittedly there has been a steady increase in the number of trout fish farms, but the main aim of many of them is to re-stock lakes and reservoirs, primarily for sporting purposes. Angling is now a leading pastime. In addition, as a counter to diminishing natural supplies caused by over-fishing, serious attempts are beginning to be made to farm marine fish domestically, sometimes using the warm-water effluent from nuclear power plants. Fish farming is, however, still in its infancy. Perhaps now, because of political and economic pressure, progress in the development of commercial fish farming to produce food will be speeded up. In the meantime, you can make your own worthwhile contribution to fish farming, and at the same time save money, in your own back garden.

1 Rules and Regulations

When you embark on a new venture, such as small-scale fish farming in your garden, it is always sensible to find out what rules and regulations exist and what permits, if any, are necessary.

Introducing fish into waters is controlled by the Salmon and Freshwater Fish Act 1975. Power to administrate this legislation is vested in the various local Water Authorities who, under section 30, can give or refuse permission for fish to be moved from one water to another.

Under this Act anybody who wishes to move fish within a Water Authority area, or between two such areas, is required to apply to do so on the appropriate form, and obtain written permission. The form is quite a simple one—your Water Authority will provide it on request. Typically you have to give the location of the water into which the fish are to be introduced, that of the water (i.e. the supplier's farm) from which the fish are to be transferred, and the date on which it is intended to do so. You also have to say whether there is any overflow from the receiving pond into a ditch or watercourse. You must state whether or not the water providing the fish is currently subject to any restriction order imposed by the Ministry of Agriculture, Fisheries and Food under the Diseases of Fish Act 1937. If the answer is 'yes', you must inform the Authority whether arrangements have been made for quarantine and inspection. You will have to contact your supplier about this information. All this may sound rather formidable at first, but if a reputable fish farmer is being dealt with the matter becomes a formality.

Even though you are stocking your pool for private use, you must still, by law, observe the close season, i.e. the period during which it is illegal to remove fish from waters by any means. The dates are determined either by statute or by local by-laws, and your local Water Authority must provide the dates of the close period as the regulations vary from area to area. As examples, the annual close season for coarse fish (i.e. carp, tench etc.) in the area administered by the Thames Water Authority is from 14 March to 16 June, while that for brown trout runs from 30 September to 1 April. At the present time there is no close season for rainbow trout. Currently there are, however, new by-laws in draft form awaiting confirmation from the Minister of Agriculture, Fisheries and Food. If these by-laws are approved, the close season for rainbow trout in enclosed waters will be the period between 30 November and 15 March.

Water supply

As it is necessary to refill a pool from time to time and to replace the water lost through evaporation, it is important to find out the regulations of the local Water Authority regarding supply. You will probably have to pay a little extra in water rates in certain areas, and sometimes the Water Board will insist on water being metered, particularly if the pond is greater than a stipulated volume. This means that you may have to pay for the installation of a water-meter and thereafter for the amount of water used instead of paying a fixed water rate. When a water meter is installed, it usually has to cover the whole of the domestic supply. A rental charge is usually made for the meter.

Electricity supply

In order to raise fish in a garden pool, you will need some sort of pump or aerator, and for this you must have an electricity cable running from the house to the vicinity of the pool. Unless you are completely qualified to lay this cable yourself, it is advisable to employ a professional electrician to carry out the work, because outside wiring can be damaged by weather conditions unless the right precautions are taken and the proper equipment used. In any case, irrespective of whether the work is done professionally or as a do-it-yourself job, it is important to know if there are any particular regulations and recommendations covering such an installation. It is wise to discuss the question with your local Electricity Board before starting.

The Local Authority

Just in case there are any by-laws covering the making of a pool for fish farming, consult your Local Authority. There probably won't be any, but it is best to make sure.

2 Adapting and Constructing Pools

What makes a suitable pool?

Size (surface area of water). In theory, fish can be raised in a pool with a very small surface area, but too many fish in a pond will lead to stunted growth and possibly disease. The recommended number of fish is one per 2 sq. ft (18.60 sq. cm) of surface. The minimum size of pool suitable for an average family is 95 sq. ft (8.85 sq. m), which would have a diameter of about 11 ft (3.3 m). This size pool should be able to sustain 45–47 rainbow trout or carp.

Depth. Fish pools should not be too deep. This may sound odd when plenty of rivers and natural ponds are very deep indeed yet are well stocked with fish. The problem with too deep a domestic pool is that it becomes difficult to remove the rotting organic material which accumulates. A satisfactory pond should not be less than 15 ins (37.5 cm) deep because otherwise ice might penetrate to the fish during winter. Also, if the water is very shallow, sharp changes in temperature are likely. If a pond is too shallow, the fish may leap out on hot summer days. An ideal maximum depth for a garden pool is 3 ft (90 cm).

Situation. Ideally a fishpond, like a herb garden, should be positioned for convenience near the kitchen. If you have children it is inadvisable to have the pond any distance from the house and out of sight.

The pool should be situated in the sun and in the open. It should not be in the shade and certainly not underneath trees. Although a little shade on a pool is welcomed to provide a cool spot on hot days, it is far better to create shade by growing on the margin some bog plants 2 ft (60 cm) high. If pools are in the shade of trees, leaves and debris fall into the pond and form a layer of decaying organic matter on the bottom which produces obnoxious gases. The fish become sickly and sometimes diseased—in fact, such decaying matter accounts for most of the losses in stock. Another objection to siting ponds under trees, especially concrete ponds, is that the tree roots can undermine the structure and cause cracking.

It is often claimed that pools completely exposed to the sun for the greater part of the day turn green because of the rapid growth of algae, but in fact just as much algae is produced in a pond which is in the shade. Algae thrives on two things—mineral salts dissolved in the water, and light. The way to combat it is to introduce into the pond natural competition in the shape of other plants that will steal the algae's food and light.

A pond is best if it is not exposed too much to severe frost. Also, in a very windy position the surface of the water can become seriously disturbed and most water plants do not like this. Winds can also give rise to rapid evaporation of the water calling for frequent replenishments. If your pool is in a very open and exposed position, some form of wind break which does not cast a permanent shadow on the pool, might be advantageous.

Adapting an existing pool

A number of gardens already have pools which quite possibly could be used for fish farming. The first question to ask is whether the pool is suitable. The best sort of pool is a natural pond, which is usually something quite permanent, created by the nature of the soil or subsoil—usually heavy clay, chalk or limestone. Often there is no difficulty in maintaining the level of the water, except possibly in a severe drought, because it is supplied by natural drainage from the surrounding ground, by an underground spring or by means of a stream flowing through the garden. Providing the input of water is approximately the same as the output, the water level will keep reasonably steady.

Emptying the pool

Unless a pool has been regularly maintained, it will probably need some attention before it is stocked with fish.

The first thing to do, irrespective of the material of which it is constructed, is to empty it so that it can be cleaned and inspected. The easiest way to do this is to pump out the water into a nearby soakaway or clean-water drain. If there is a pump for circulating water through a fountain or over a cascade, this can be easy; disconnect the pipeline, connect it to a hose leading to a clean-water drain, and start the pump. If this is not possible, a pump can often be hired from a local plant hire company or a builder. This might be necessary if you are dealing with a natural pond supplied by a spring or a stream; this type of pond will require continuous pumping, especially if a lot of work needs to be carried out. It is also possible to buy a pump that operates on an electric hand drill; this is excellent for emptying pools (see Fig. 1). This kind of pump will handle up to 200 gallons (910 litres) per hour and is simple to operate—all that has to be done is to connect ½-in. (1.25 cm) hoses to the inlet and outlet and push the spindle into the chuck of the drill. It should not, however, be used continuously.

The level of my own pool is about 3 ft (90 cm) above that of a conveniently positioned clean-water drain. The pool can be readily emptied by connecting a hose to a tap, placing the further end well under the

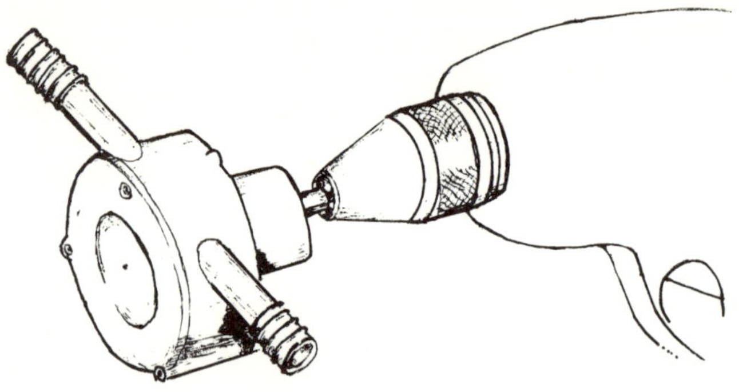

Fig. 1 Electric drill pump which operates with any electric hand drill having a speed range between 800 and 3000 rpm.

surface of the water, and turning on the tap. After the water has been running freely for about half a minute, I disconnect the tap end of the hose, place my thumb on the aperture and move it to the drain, keeping the hose as near to the ground as possible. As soon as I remove my thumb the water in the pool siphons until the pool is empty. If you have any variations in levels in your garden, check to see whether you can do the same.

Failing any of the methods already described, there is no alternative but to embark on the arduous task of emptying the pool with a bucket. If you have a portable water butt available, it makes the job easier.

Cleaning the pool

The important next step is to clean the pool. Firstly, all the existing growth must be thinned, removing as far as possible all the dying or rotting tissues. All decaying matter, such as dead leaves, twigs and other organic material, must go. You should also try to remove the cause—consider whether any nearby offending trees or large shrubs could be cut down or pruned without detriment to the garden. Certainly if such trees are overhanging the pond it might be essential to do so, to give more light to the water and make it a healthier place for the fish.

Repairs

The next step is to repair the pool, if necessary. If it leaks continuously, it might be due to a hole, crack or tear in the liner. A constant leak in a natural or concrete pool, with nothing visible to indicate an escape point, could mean that the surface is porous.

A natural pool is likely to be lined with a thick layer of clay. To make this good you must reline it with fresh clay, puddling it in well. The very existence of a natural pool usually means that the soil in the garden or the locality is clay, and there should be no difficulty in obtaining a supply. Failing this the pond can be concreted over or lined with butyl rubber.

Established man-made pools, especially in older gardens, are usually constructed in concrete. The faults you may find are cracks—sometimes hair cracks, sometimes wider openings—that have arisen mainly from earth movement. Sometimes the top layers will be found to have broken away over a substantial area. To repair this kind of pool, cut open all the narrow cracks a little wider. Then fill them with a cement mixture consisting of two parts sand and one part cement, which can have some waterproofing compound added to it to increase its efficacy. It is very important to mix this rendering coat very thoroughly, and it is best mixed in relatively small quantities. If this is done, a good seal can usually be achieved even without the addition of a waterproofing powder.

If the rendered surface is actually broken anywhere, the immediately surrounding sound area should be lightly chipped to make a key. The whole area should then be rendered with the mixture described above. A rather better seal can be obtained by covering the broken part, and the good part for a little distance surrounding it, with a small sheet of polythene or PVC, held temporarily in position by three or four small blobs of rapid-hardening epoxy resin or similar adhesive. The sheet is then rendered fairly generously with a cement and sand mixture.

If, when you empty a cement-lined pool, you find no visible signs of a leak, yet when the pool is filled water seeps away at a rate faster than you would expect from evaporation, the cement rendering is most probably porous. This is usually caused by the sand and cement not being thoroughly mixed in the first place, so that throughout the surface there are areas that are not waterproof. If this is the case, prepare a well-mixed slurry from two parts sand, one part cement and enough water to make the mixture just workable with a paintbrush. Paint this carefully all over the surface of the pool. (It is also a good idea to give a similar coating to any surface treated in the various ways described above.) The cement should be allowed to dry slowly, taking one week to ten days. You can slow down the natural drying rate by covering the newly rendered surface with a sheet of medium-thick polythene. Slow drying helps to make the cement coating waterproof.

Alternatively, cracked and porous concrete pools can be treated with a proprietary plastic sealing paint. It is, however, essential to make sure that any bituminous material previously painted on to the surface is thoroughly removed. These plastic preparations should be applied when the weather is dry and warm. They are usually sold in blue and stone colours, which give a pleasing appearance to a fish pool.

If the garden is not very old it is quite possible that any existing pond will have a plastic liner. The most common materials used for liners are polythene, butyl synthetic rubber and PVC. Of these, a synthetic rubber liner is probably the easiest to repair. It can be repaired *in situ* with a repair kit, available from do-it-yourself and liner suppliers. The kit contains the correct adhesive, patching tape and instructions. A PVC liner can also be repaired on site by means of a patch of the same material and an adhesive such as Bostik No. 1 Clear. There is really no satisfactory method of repairing polythene if a puncture is found. Replace it with one of the better sheetings.

Extending an old pool

It is difficult to make an existing pool larger unless it is a natural pool, i.e. one in which the water is retained because of the nature of the soil in the garden. It can be done, however, to some extent with a pool that has a plastic liner, but it is better to anticipate the possibility of needing an extension when it is first made.

With a natural pool, empty it, then dig out more soil until the required shape and total new size has been reached. After that the new walls and bottom will have to be coated and puddled with a thick layer of clay (see page 25). Particular attention must be given to making sure there is a good, sound bond at the joint between the old and new surface areas.

There are three possible ways to extend lined pools. The first is to provide, when the pool is first made, a larger-sized liner than is actually needed and to fold the surplus sheeting carefully under the coping stones. When the pool is to be made larger, it is emptied, the necessary extra soil excavated, and the plastic liner put into position. During the operation the liner should be folded back and great care taken to protect it from damage.

The other two methods of extending lined pools apply only to pools with butyl synthetic rubber liners and, again, some provision must be made for doing so when the pool is originally fabricated. Firstly, the supplier can vulcanize an additional section on to the liner at his factory, and secondly, although it is rather less satisfactory, extra sheeting can be joined on to the original on site by means of joining tape and adhesive. It is necessary in both cases to provide for easy access to the liner when the work is to be carried out. So the possibility of extension has to be anticipated in the first place.

PVC and polythene liners cannot be extended after installation.

Making a deep pool shallower

If your pool is deeper than 3 ft (90 cm), place a layer of gravel on the bottom to raise it to the correct height.

Constructing a new pool

If you have not got a suitable pool already, you will have to construct one. In addition to its more practical aspect of contributing food, it will also enhance the beauty of your garden.

Size and shape

In the first place, from an aesthetic viewpoint, it is important to keep the shape a good, open, unsophisticated one. A small amount of irregular curvature will create an aspect of informality, but it should not be extravagant. The more unusual shapes, such as figure-of-eight, wiggly outlines and so on, might seem attractive on paper, but on the ground these shapes will not be so noticeable. Basically square and circular outlines should be adhered to in formal surroundings, and irregular broad ellipses and ovoids in more informal ones. From a design point of view it is important to make sure that the size is proportionate to the overall area of the garden.

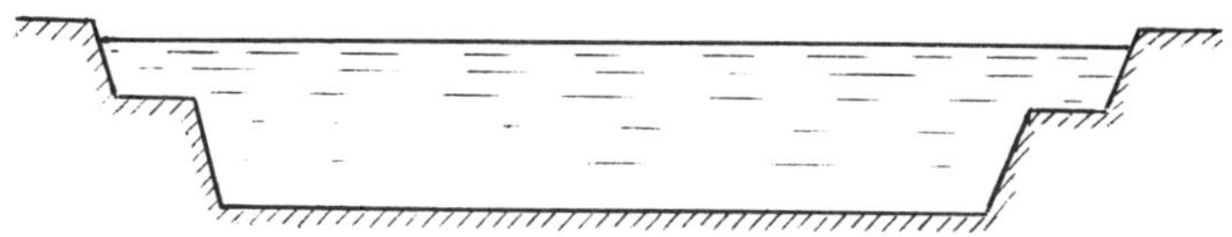

Fig. 2 Cross-section of an ideal pool. This shape gives a high volume of water in relation to the surface area.

The sectional shape of a pool (i.e. the shape that would be seen if the pool was cut in half down the middle and looked at horizontally) is an important consideration as far as plant and fish life are concerned. It must be of such a form that the pool contains a high volume of water in relation to its surface area (see Fig. 2). A small volume of water gives abnormally high growth of algae and the water remains permanently green; it is also subject to rapid fluctuations in temperature and, of course, provides living accommodation for a smaller number of fish.

Lastly, check that the area where you propose to site your pool is level. If it is not, appropriate steps must be taken to rectify this fault.

Making a natural pool

The luckiest people are those who have clay soil in their garden, because they have to do little more than dig a hole and fill it up with water from a hose. Soon teeming with insect and plant life, it becomes an ideal home for fish. At first it is often necessary to top up this type of pond fairly frequently, depending on the porosity of the soil structure, but, if fish are present, after two or three years their droppings will seal the pond.

To make a natural pool dig out soil to the desired shape, size and depth. Make the bottom very smooth and tidy, and ram it down quite hard. The best tool to use for this is a pavior's iron maul, which might possibly be borrowed from a local builder. This surface should then be covered with about $\frac{1}{2}$ in. (13 mm) of silver sand in which the fish will nose about.

If you have a stream flowing through your garden, you can dam it and widen it to form a pool with the stream entering at one side and leaving at the other, maintaining a constant level. There is no better environment for fish and water plants. If this kind of pool is to be stocked with rainbow trout, buy a non-migratory stock.

Artificial pools

So much for the lucky ones. But most people who want to farm fish in their gardens have to make an artificial pool. Usually they want it to be something attractive, which does credit to the rest of the garden. This, however, is not essential, because fresh-water fish can be successfully raised in a sizable tank, possibly picked up at a scrap-iron merchant's yard. It should be buried in the ground for most or all of its depth. Quite possibly its water supply would be derived from the rainwater guttering of the house. It would also be useful in a drought.

There are three ways in which an ornamental pool can be constructed artificially. The first is by means of a prefabricated fibreglass pool, which can be formal or informal in shape. Generally these are not made in large enough sizes to be fully satisfactory for fish farming. Perhaps their greatest disadvantage, compared with other types, is their high cost. They also have a relatively low estimated life of ten years.

Probably the best and easiest method of making a pool is to excavate a suitably-shaped and -sized hole and line it with an impermeable liner. The best of these are made of butyl synthetic rubber, which is claimed to have a life of about fifty years, providing it is not ill-used. Even if accidentally punctured it is fairly easily repaired. In addition, it is relatively cheap. There are also several types of PVC pool liners; although in most cases they are cheaper than butyl rubber liners, this is offset by their shorter life—about ten years. Finally it is possible to line a pond with polythene sheeting, but this is not satisfactory. It cannot be repaired and its estimated life is no more than two years. It is nevertheless cheap and is useful for temporary use.

The third material used for constructing a garden pool is concrete, which can, if properly done, make a good, fairly permanent pool. It is, however, not entirely reliable. In the first place, its ingredients must be thoroughly mixed to make sure that it is waterproof. A concrete pool is subject to frost damage caused by the pressure of ice, formed when the water freezes, fracturing the walls. They may also be fractured by earth movement. It is a laborious task to use concrete to make a pool. Its estimated life is authoritatively given as twenty years yet it has been known to break down after a year. The cost of making a concrete pool is almost as high as using a butyl rubber liner.

Making a pool with a liner

Before you buy an impermeable liner you must ascertain the size that it has to be. This can be done while the pool is still just a sketch on a piece of paper, providing its dimensions are finalized. The size required is the overall length + twice the maximum depth × the overall width + twice the maximum depth. This formula will allow for any overlap at the top if the pool is dug with the sides sloping inwards at an angle of 20° to the vertical, or 3 ins (7.5 cm) in for every 9 ins (22.5 cm) down, as recommended. The

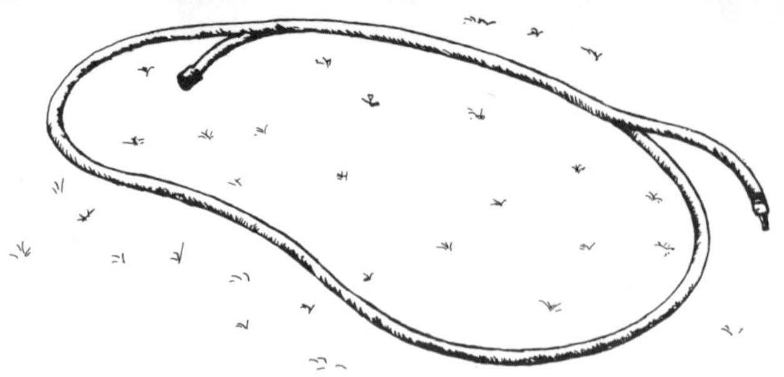

Fig. 3 Hosepipe laid out on the ground to mark the shape and size to be cut.

inward-sloping side and the slight stretch of the material will allow adequately for any overlap.

1. Outline the desired shape and size on the site, which must be level, using a rope or a hosepipe. (Store your hosepipe in a warm place overnight to ensure that it is absolutely flexible and can take up any shape required.) (Fig. 3.)
2. Cut out the outline with an edging iron or a sharp spade.
3. Cut the sides of the pool with an inward slope of 3 ins (7.5 cm) for every 9 ins (22.5 cm) down. If you want a marginal shelf, remove the soil for a vertical depth of 9 ins (22.5 cm) over the whole area and then start cutting and removing the soil to the predetermined depth at a point 9 ins (22.5 cm) horizontally from the base of the initial cut, to form a 9-in. (22.5-cm) wide shelf all around the circumference of the hole. (See Fig. 4.)

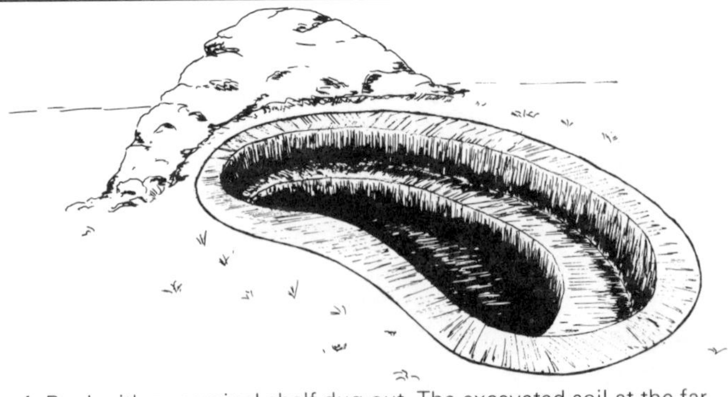

Fig. 4 Pool with a marginal shelf dug out. The excavated soil at the far end is for a rockery.

Fig. 5 The liner in position and weighted down at the corners.

4. If you want to make a rockery and install waterfalls in conjunction with the pool, place the excavated soil in a suitable position near its boundary. (Fig. 4.)
5. See that the top edge of the pool is level right the way round. At the same time check that all the dimensions are correct.
6. Remove all sharp protrusions and stones that are likely to cut the liner. Then work sand into the walls and bottom by hand. If the pool has been dug on very stony ground, put down a layer of newspaper or polythene. The surface should be neat and smooth before the liner is placed into position.
7. Drape the liner with its full area uniformly distributed in the hole and weight it down at the corners with bricks or large stones. (Fig. 5.)
8. Start filling the pool with water. Ease off the weights to allow the liner to take up snugly the shape of the cavity. If the pool is a formal rectangular one, fold the liner into the corners before filling.
9. When the pool is full reduce the width of the overlap to about 4 ins (10 cm) overall by trimming

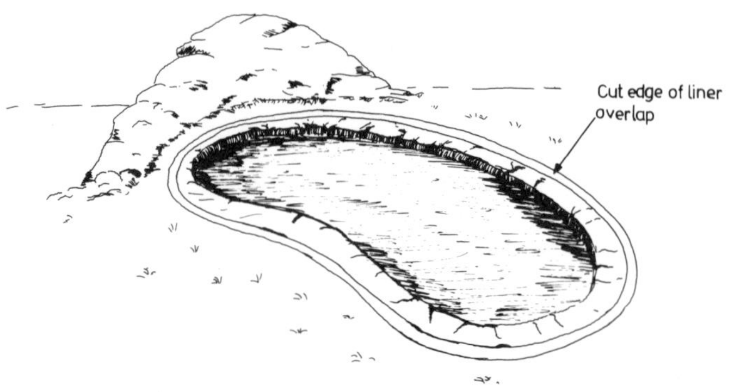

Fig. 6 The filled pool with surplus liner material trimmed off.

away the surplus material. If necessary this can be secured temporarily by driving 4-in. (10-cm) nails through the flap. (Fig. 6.)

10. Pave the surrounds of the pond with crazy or random stones anchored with blobs of mortar mix (3 parts sand to 1 part cement), so that they do not move, and so secure the liner. Each stone should overlap the edge of the pool by 2 ins (5 cm) to hide the liner. (Fig. 7.)

Constructing a concrete pool

Since the introduction of the more satisfactory impermeable liners already described, concrete is not

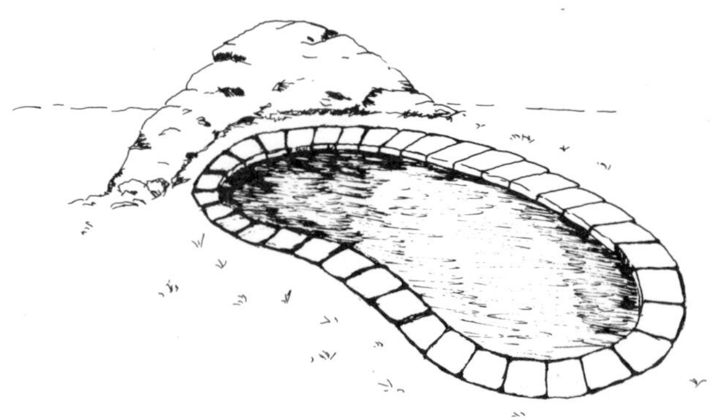

Fig. 7 The liner has now been secured with paving stones round the edge.

used for pools as much as it used to be. There is, however, still one situation in which it is useful: in building a raised pool in which the water is filled into a square or round enclosure with a wall, which stands about 2 ft 6 ins (75 cm) above ground level. This type of pool is particularly suitable for installing where there are young children. In addition, a pool of this nature is very suitable for a formal paved area, where considerable interest in design can be created through varying the levels of the features.

1. Mark out the outline of the base of the pool on the ground.
2. Excavate soil from an area 5 ins (12.5 cm) wider all round than the base, to a depth of 1 ft (30 cm).
3. Make a good foundation for the pool by putting at the bottom of this hole a 6-in. (15-cm) layer of brick rubble, firmly rammed down, and on top of this a layer of concrete 5 ins (12.5 cm) thick, consisting of three parts coarse—1-in. (2.5-cm) down—aggregate, two parts sand and one part cement. These quantities should be measured accurately by volume. The ingredients must be very thoroughly mixed first while dry and then with water to give an even, smooth consistency which is just pourable. This thorough mixing is essential to ensure that the concrete is completely waterproof. A waterproofing compound may be added, following the manufacturer's instructions. All the concrete used in constructing the pool should be allowed to harden slowly, by covering it with sacking or polythene for about a week.
4. When this concrete has hardened, erect double hardboard shuttering, appropriately supported with timbers, centrally on this foundation so that there is a 5-in. (12.5-cm) margin all round. The inner and outer walls of this shuttering should be 5 ins (12.5 cm) apart.

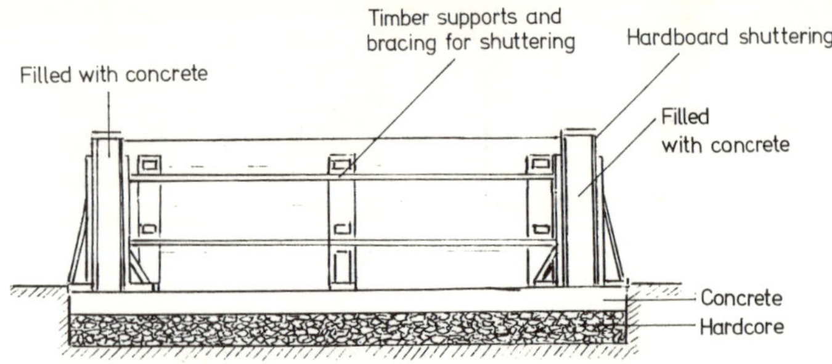

Fig. 8 Cross-section of a raised pool, showing foundations, timber supports and hardboard shuttering.

5. The space between the shuttering is then filled with a concrete mixture of the same composition as that used for the foundation. (Fig. 8.)
6. When this concrete has been allowed to harden slowly, the shuttering is removed.
7. The inner surface of the pool is rendered with a 1-in. (2.5-cm) thick coating of mortar, composed of two parts by volume of sharp sand and one part cement. To ensure that the surfaces are completely waterproof, mix these ingredients thoroughly. (Fig. 9.)
8. An alternative to cement rendering is to use a butyl synthetic rubber liner, which will ultimately be fixed by the coping stones on top of the walls.
9. If you want to make the raised pool attractive on the outside, the rough concrete wall could be faced with brick or stone. (Fig. 10.)
10. Finally, a brick or stone coping with an overhang on either side of about $1\frac{1}{4}$ ins (3.1 cm) should be

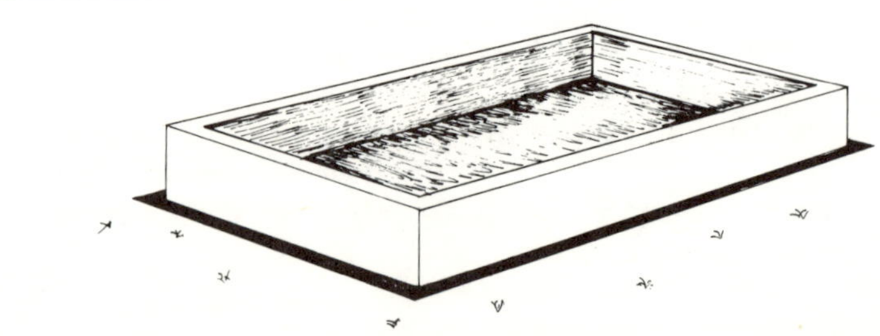
Fig. 9 Concrete pool with shuttering removed and rendering applied.

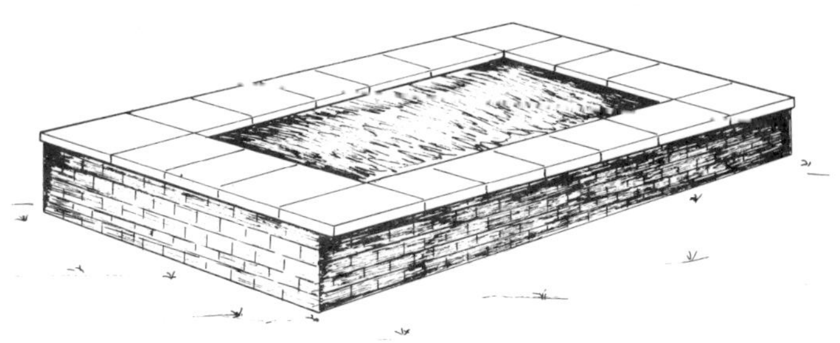

Fig. 10 Concrete sides faced with bricks, and coping in position round the edges of the pool.

built on top of the wall to give the pool a finished appearance, and also to fix the liner if you have used one. (Fig. 10.)

Although perhaps a little more difficult, especially in making the shuttering, a round raised pool can also be made. It is well suited to having a fountain or aerator positioned centrally.

As concrete yields free lime which is soluble in water, a concrete-lined pool should not be filled and stocked with fish or plants without further treatment, because neither will tolerate it. You should paint the surface with a proprietary compound in accordance with the maker's instructions, or with five coats of a solution of colourless magnesium silicofluoride or sodium silicate (waterglass). If there is plenty of time, the simpler and possibly more reliable treatment is to fill the pond with water and allow it to stand for about two weeks. If this is repeated three times the pool should be safe. If the pool is lined, none of this has to be done, which, of course, is a great advantage.

A temporary pool

Polythene, though not very satisfactory for lining a permanent pool, is nevertheless very useful for making a small temporary pool, since this material is cheap and easily rolled up for storage when not in use. It is particularly useful for making a pool to be used temporarily for housing fish and aquatic plants, such as water lilies, when the main pool is being cleaned out or repaired. In addition when fresh-water fish, such as carp, breed in captivity, the tiny young fish should be removed from the main pool and kept separately until they are 2–3 ins (5–7.5 cm) long. If this is not done, some of the larger fish may eat them. Wooden barrels or metal drums are often used for this purpose, but unlike polythene sheeting they need a lot of storage space when not in use, and this is seldom available in today's small gardens. A temporary pool could also be used to house any fish that show signs of being unwell.

Safety of children and pets

Perhaps the best way of dealing with the problem of safeguarding children and pets is to cover the whole surface of the pond very securely with strong, fine nylon mesh which, when tightly secured round the edges of the pond and supported on timber cross members, can withstand the weight of several small children. A thick planting of marginal plants would deter young children and animals from stepping on to the netting. The mesh will not spoil the appearance of the pool because aquatic plants will grow through it. It will not affect the fish at all.

Another method is to fence the pool in around its entire perimeter to a height of up to 3 ft (90 cm), with either a green plastic-covered wire fence or similar 2-in. (5-cm) mesh chain-link fencing supported on metal or wooden posts. The latter is less prominent. Admittedly both kinds may look unattractive, but you can soften the effect by planting groups of herbaceous perennials at intervals on the outside of the fence.

If your pool is a raised one, say 2 ft 6 ins (75 cm)–3 ft (90 cm) above the ground, there is no risk to small children.

3 Waterfalls and Watercourses

The need for oxygen

In order to live, fish need sufficient oxygen. Among fresh-water fish the amount needed varies from fish to fish. Regarding the kinds likely to be kept in a garden pool, rainbow trout need water that is teeming with oxygen, whereas carp and tench are happy with much less. Fish obtain their oxygen largely from the air and turbulent water absorbs air in considerable quantities. In the course of a river's run from its source, probably high up in the mountains, to the sea, it passes through various stages creating a number of different environments that are optimum for a range of fish. High up, for instance, the stream rushes down rocky slopes. As it sweeps by these obstacles the water leaps and surges and absorbs copious quantities of oxygen from the air. Later, as the bottom of the watercourse changes in nature from hard rock to soft rock, the soft rock is worn away over the years and this creates a vertical fall. As the water tumbles over the edge of this waterfall, it is divided into numerous small streams, which, because of the greater exposure to the air, absorb still more oxygen. As each watery spout falls into the pool below it carries quantities of air to the bottom of the water, where once again life-giving oxygen is released into the water. As the slope of the river bed levels out a little lower down and the stream widens and becomes calmer, fish and water plants, which also provide oxygen, begin to appear. Soon, while the water is still cool and lively with oxygen, brown trout are found; a little lower down, as the oxygen-laden waters become a little warmer, rainbow trout abound. Still later, as the bed of the river becomes more level, the stream runs more slowly, the rate of absorption of oxygen falls and the waters begin to warm up, fish such as carp and tench, which can live with less oxygen, appear. These waters also produce an abundance of vegetation, which provides food for them.

If fish such as rainbow trout are to be raised successfully in a garden, the first essential is to make sure that the water is teeming with oxygen. For other fish—carp, for example—rather less oxygen is necessary. In fact, they can cope in still water. One way of ensuring that enough oxygen is available in the water of a man-made pool is to simulate as far as possible what happens in Nature. More especially in an informal pool this can be readily done by introducing a watercourse with waterfalls and rapids. Apart from its practical value, such a layout would make a superb focal point for the garden. The extent to which water can be adequately oxygenated in this manner depends, of course, on the space available. If the pool complex is relatively small and there can only be one, or at most two, relatively narrow waterfalls to maintain the correct proportion, it might be necessary to install an informal fountain, or better still a waterspout, which rises some feet into the air and then splashes vigorously on the surface of the water to provide oxygen. In a formal pool in a paved area, simulating Nature would be unsuitable and you would have to install a fountain. Fountains and waterspouts need a pump to keep the water in circulation. Fountains, mechanical aerators and pumps are discussed in Chapter 4. This chapter describes methods of constructing a watercourse with artificial rapids and waterfalls as a means of providing oxygen.

Building a watercourse

The first stage is to form a mound 2–3 ft (60–90 cm)

high at the side or back of the pool. If you are constructing a pond for the first time this is relatively easy, because the soil that has been excavated can be used. On the other hand, if an existing pond is being dealt with, it involves bringing soil from elsewhere. Providing the site is well drained and a 6-in. (15-cm) or so layer of good soil covers the mound, the quality of this imported material can be quite indifferent as long as it is reasonably porous. Apart from providing space for constructing secondary pools and waterfalls, this mound is used to make a rockery which also disguises the artificial material that has to be used. Naturally, the longer the watercourse can be and the greater the height of the waterfall or waterfalls, the larger will be the amount of oxygen absorbed.

Now you must install a pipeline underneath the mound so that the water can be pumped from the bottom pool up to the top of the highest waterfall. (Pumps are described in Chapter 4.)

As with a pool, there are three types of materials that can be used for making a watercourse—fibreglass, butyl synthetic rubber and PVC liners, and concrete. Basically, whichever method is used, the first step is to make the foundation in the mound of soil. This should preferably follow a slightly twisted line, forming a series of steps with treads and risers of sizes and heights that suit the features you want to incorporate (see Fig. 11). A

Fig. 11 Starting a watercourse. A flowing line of steps cut into the mound.

number of combinations can be used, but it is quite effective to have a small rock pool at the summit followed by at least one waterfall, and a final one as the water enters the main fish pool. This, however, depends entirely on the height of the rockery mound. A vertical drop of 9 ins (22.5 cm) gives a very good splash and the supply of oxygen in the water will be renewed. If the height of the watercourse above the level of the water in the pool is 3 ft (90 cm), it is feasible, therefore, to have up to four waterfalls. (Fig. 12.)

There is little doubt that the easiest way to construct a watercourse is to use preformed units of fibreglass or plastic, which are light to handle, or, more substantially, of concrete. Having made suitable bases on the treads all that is necessary is to bed them into the soil and flank them with rockery stones. Try

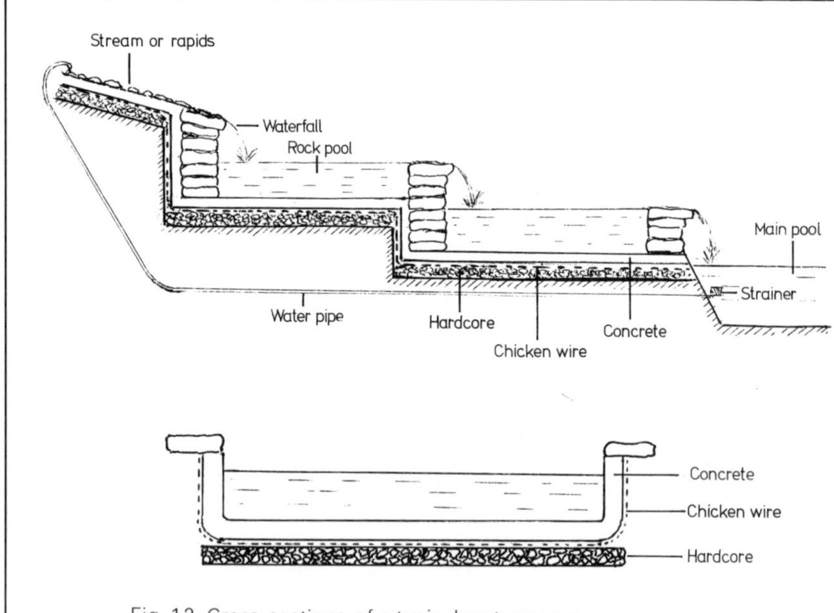

Fig. 12 Cross-sections of a typical watercourse.

to make the effect look natural. It is possible to purchase complete waterfall kits, which include a pump and even a collection of aquatic plants and planting crates.

When constructing a watercourse it is vital to make sure that it is watertight, because any leakage will result in a fall in the level in the fish pool. As much water must return to it as is pumped out. To avoid leakages ram down the bottom soil so tightly that there is no risk of cracks or distortions.

No doubt the most satisfactory way of making a watertight watercourse is to use concrete to form a base over the whole length from the top right down to the edge of the main pool at the bottom, the risers of the steps cut out of the mound of soil, and the sides. To do this successfully the concrete layer should be at least 3 ins (7.5 cm) thick. It should consist of three parts by volume of coarse—1 in. (2.5 cm) down—aggregate, two parts of sand and one part of cement. All these ingredients should be mixed together while dry, and then thoroughly mixed with enough water to produce an even, smooth, just pourable consistency. This will ensure that it is waterproof although as an extra precaution a waterproofing compound can also be mixed in. This, however, is not an absolute necessity, providing the ingredients are mixed well. When made, the concrete surface should be covered with sheets of polythene weighted down with bricks, so that the concrete hardens slowly over about a week.

The first thing to do is to make the required number of steps in the mound, according to the number of waterfalls you want. On the treads of these steps a layer of hardcore or brick rubble should be pounded down to give a 3-in. (7.5-cm) layer that will form a sound foundation. When digging out the steps in the mound, you should make allowance for these layers of hardcore and concrete (see below). After putting down the layer of hardcore, the concrete should be reinforced by chicken wire on the treads, up the sides of the watercourse, and on the risers of the treads. This should be covered with the 3-in. (7.5-cm) layer of concrete. The most convenient way of doing this against the treads is to build a stone wall from the step below 3 ins (7.5 cm) away from the step and of such a height that it forms the face of the waterfall and a retaining wall for the pool above it. To give this wall the natural effect of a waterfall, the top stone should be allowed to overhang the lower stones by about 2 ins (5 cm). The concrete is poured in between the riser and the wall, making sure that there is a waterproof joint between it and that on the tread. A layer of concrete of the same thickness should be built up on the sides of the watercourse.

The stones used for making the waterfalls should be bonded together with mortar, made by carefully mixing together two parts by volume of sand and one part cement, with enough water added to make a workable mixture. As before, a waterproofing compound may be added if desired. It is important for the top stone of the wall to overhang about 2 ins (5 cm) because this makes the water flow over the lip of the waterfall in a broad, sometimes divided, film producing a good splash. On no account must the water be allowed to trickle slowly down the face of the rocks otherwise oxygenation of the water will be poor. From an aesthetic point of view, allowing the top stone of a waterfall to overhang gives a more natural appearance, suggesting the effect of the erosion of the lower rock face that normally occurs in a waterfall.

If possible a watercourse should include a length of sloping stream containing small rocks fixed by cement to the base of the channel so that their upper surfaces are either just level with, or just above, the level of the flowing water. This creates turbulence, which produces further absorption of oxygen.

Various types of stones may be used in creating a watercourse, but in terms of beauty nothing can surpass the water-worn, flat, rounded ones found in mountain areas. Another possibility is sea-washed, clean beach pebbles of varying sizes. Failing both of these, you can resort to the more easily obtainable, weathered Westmorland rockery stone.

A watercourse constructed in concrete must be treated with a chemical compound to neutralize the lime in the concrete. If pumps are installed an alternative, but lengthier, method is to fill the main pool with water and allow it to circulate continuously for about a fortnight. Then the system should be emptied and re-filled. This should be repeated at least twice more. After this, fish and plants can be introduced.

The alternative to a concrete shell is to line the watercourse with butyl rubber sheeting, which is by far the most suitable of this type of structural material. However, it is not so easy to use as concrete, although much less work is involved. This is because it is not so adaptable to forming a pouring lip from one rock pool to another. It has been found that it is rather more satisfactory to use separate pieces, instead of one whole length. One supplier offers liners of various dimensions suitable for different stages in a watercourse, e.g. rock pools and short and long streams. Fig. 13 shows cross-sections of a watercourse in which the underlying waterproof membrane is of butyl synthetic rubber. Particular care must be taken to see that the end of the sheet of butyl rubber at the head of each waterfall is arranged so that it forms a satisfactory pouring lip. This can be achieved by placing moderately small rocks on it just before the fall, so that it is held in position.

Cut steps out of the mound as in the case of the concrete watercourse. The bottom pool is formed by

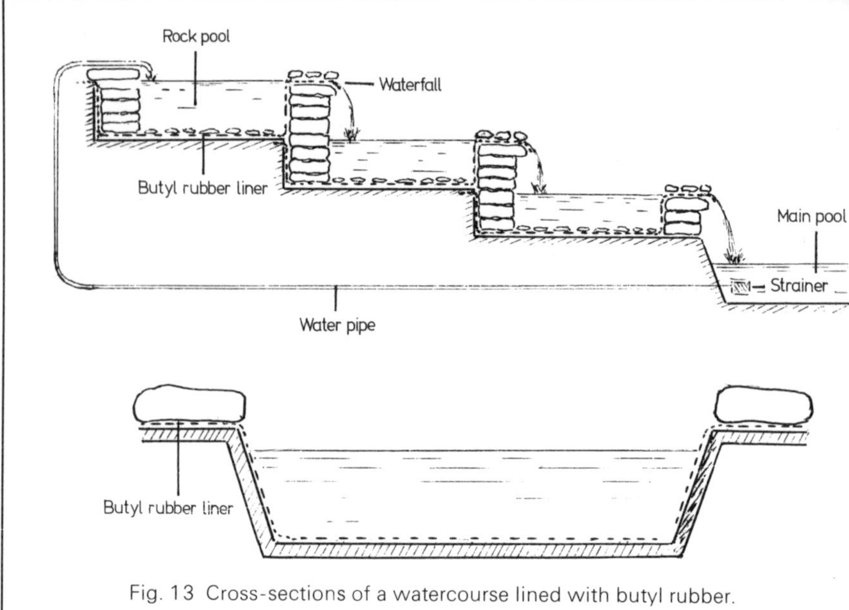

Fig. 13 Cross-sections of a watercourse lined with butyl rubber.

building a parapet wall on the edge of the main pool (Fig. 13). Line the basin created between this wall and the first riser with a sheet of butyl rubber and secure it at the sides and front with rocks at the top. Pack soil between the liner and the wall to prevent the liner being punctured. Next, construct a wall against the rear riser, securing the liner—the wall should be of such a height that it forms another rock pool with the next higher riser. Line this basin similarly. Continue in this way until you reach the top of the mound.

For rainbow trout you will also need to install a water spout in the pond.

4 Oxygenators, Pumps and Filters

The importance of having adequate quantities of dissolved oxygen in the water in a fishpond has been emphasized throughout this book. There are two means of producing oxygen for this purpose. The first is by planting submerged oxygenating plants in the pool (see Chapter 5), and the second is to expose water in thin films and spouts to the air so that the oxygen in the air can dissolve freely in the water. In Chapter 3 waterfalls were discussed as one means of achieving this. This chapter, however, discusses fountains, other mechanical types of oxygenator and pumps.

Fountains

Fountains suit formal pools in which a watercourse with waterfalls would be quite inappropriate. Water playing on the surface of the water of a pool for long periods from the jet of a fountain can be an excellent oxygenator. A fountain can also make a most striking focal point in the garden. The water jets can emerge in many patterns, but apart from the decorative effect, the shapes to choose are those that divide the water into as many tiny streams as possible, so that it gets maximum exposure to the air and masses of drops fall on to the surface of water, each one carrying air with it. It is important that the jets reach the maximum height possible, so that they fall on to the water's surface with as much pressure as possible.

Never allow a fountain to play constantly on any water plants.

Low energy aerator

A recently patented innovation for oxygenating water

A low energy aerator in operation.

in a fish pool is a low energy aerator (see illustration). The suppliers state that it is possible to harvest up to 100 lb (45.4 kg) of trout annually from a pool that is 9–12 ft (2.7–3.6 m) in diameter and not less than 12 ins (30 cm) in depth. It is claimed to provide a highly efficient means of oxygenating the water by allowing it to cascade out in such a manner that every droplet is full of oxygen. As a result the fish are exceptionally lively and have a very healthy appetite, which helps them to keep in good health and grow rapidly.

This aerator is equivalent to a fountain and can be installed in a fairly modest-size garden pool of the dimensions given above, and up to a surface area of about ½ acre (2024 sq. m). Its approximate overall dimensions are 18 ins (45 cm) in diameter and 11 ins (27.5 cm) in height. It is made of plastic and sits on a balsa base which floats on the surface of the water, so there is no need to build a plinth for it. It has a built-in water pump driven by an electric motor designed for continuous running throughout the day and night. It is supplied with 21 ft (6.3 m) of electric cable and arrives ready for immediate use. As it has a tendency to gyrate, the suppliers recommend that it should be stabilized with two thin nylon cords from each side of the pond.

The aerator's consumption of electricity is very low—rather less than that of two 100-watt electric light bulbs. Its voltage is 230/240 and it makes very little noise when running—about the same as is produced by a $\frac{1}{4}$-hp motor, but even this does not cause any great disturbance because it is masked by the sound of splashing water.

As the aerator needs a minimum depth of 12 ins (30 cm) of water in which to operate efficiently, it is important in ponds whose depth is at this minimum level to ensure that any loss by evaporation is regularly replaced. Although it is not absolutely necessary, in order to keep a good concentration of oxygen continuously, it is an advantage to keep the aerator running for 24 hours a day. It can be switched off at night as an economy measure. If, however, it is operated day and night, it is fairly certain that the water will not freeze in periods of frost. Nevertheless, if it is stopped and the water does freeze, as the aerator is made of resilient material it can be re-started quite easily without damage.

Pumps

Pumps are necessary in a water garden or pool to maintain waterfalls, watercourses and fountains in operation. There are two types of pump available for water circulation: one is the submersible pump, which needs very little installation, and the other is the surface pump, which needs housing outside the pool and also more elaborate plumbing. Apart from capacity, an important consideration when choosing a pump for a fish pool is whether the pump is capable of being run continuously. This is determined by the type of electric motor with which it is fitted. If it is a series-wound motor, it is only suitable for intermittent running—not exceeding six hours at a time. On the other hand, if a pump is driven by an induction-type electric motor it can be run continuously. If a pump with an induction type of motor is installed, it is recommended that a starter with an overload trip is used. This installation is essentially a job for a qualified electrician.

Apart from this consideration the other determining factors in choosing a water pump are the size of the pool, the volume of the water to be moved per hour, the height to which it has to be lifted and the number of features that have to be supplied.

Selecting the right pump

If you have any difficulty in choosing the best pump for any purpose, the best course is to seek the help of a supplier. Given the dimensions of the pool, how many fountains (giving the height of emerging jets and spread required) and waterfalls (giving the width of the pouring lip and the vertical lift, i.e. the height of the highest waterfall above the surface of the water in the pool), he will often give you advice. In fact some suppliers have a pump exchange service, under which a pump purchased that proves inadequate in its performance can be exchanged for a larger one if returned within seven days.

Submersible Pump: Although there are rather more expensive models capable of high output and high pressure suitable for large waterfalls and multiple fountain displays, a submersible pump is ideal for an average-size garden pool with one fountain and/or a waterfall. All you have to do is connect the cable, which is supplied already sealed into the pump, to the power supply outside the pond by means of the waterproof connector supplied. For safety this

A submersible pump and strainer.

connector should be covered by a paving slab. Lower the pump into the water and onto a plinth of bricks on the bottom of the pond, which will keep it clear of mud and debris. It should be at such a height that the fountain jet, which is usually attached to the pump, is just above the surface of the water. The pump, which works silently, draws water to the fountain through a strainer, which is attached to the pump. If desired, it can also draw water simultaneously to the head of the watercourse through a plastic hose via a T joint fitted into the vertical pipe leading to the fountain jet. A gate valve is fitted to each outlet to control the flow. The accessories are usually supplied as a kit with the pump.

Surface Pump: Under some circumstances, a surface pump is preferable to a submersible pump. Among these are:

(a) When the head required for a waterfall is comparatively high.
(b) Where a number of features (e.g. a waterfall and several fountains) are wanted simultaneously.

A popular surface pump kit with three outlets: one for a waterfall, two for fountains. A foot valve and strainer is shown on the right.

(c) When the demand is beyond the scope of a submersible pump.

A surface pump must be housed in a well-constructed, well-ventilated, dry chamber. It draws water, by way of a plastic tube, through a strainer situated in the pond and despatches it to the waterfalls and fountain through further plastic pipes fitted to T joints on the delivery side of the pump. The output through each is controlled by a valve. The pipes should be kept as short as possible and should be of the diameter recommended by the manufacturers.

There are two ways of housing a surface pump. The first is to keep it above the pool level in, say, an attractive stone box in keeping with its surroundings. In this case, to keep the pump primed a foot valve and a strainer must be used on the suction tube. The alternative is to accommodate the pump below the

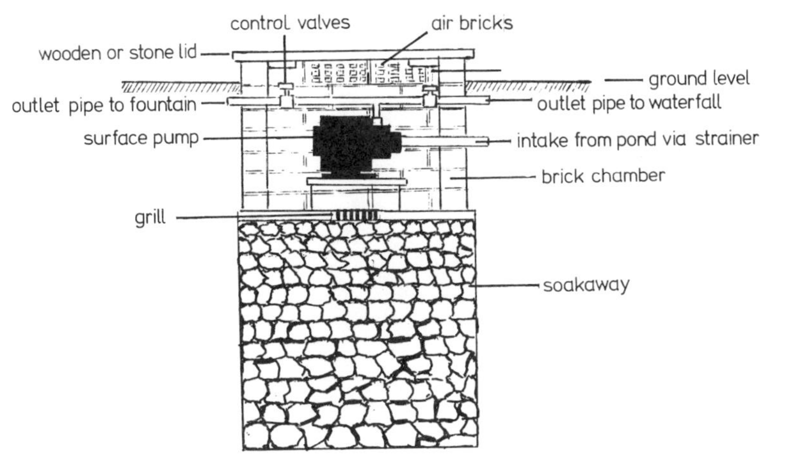

Fig. 14 Cross-section of a surface pump held in a brick sump.

water level in a concrete or brick sump built on top of a soakaway with a grill-covered drainage hole underneath the seating of the pump, so that there is no danger of flooding. In this case it is only necessary to fit a strainer, because the pump is primed by gravity. The pump must not be positioned at a height above the water level that is greater than its suction lift. (See Fig. 14.)

Filters

Recent studies have shown that it is necessary to include an efficient filter in any water-circulation system. This will remove dirt and decaying organic matter and so minimize the risk of a possible build-up of ammonia, which would be detrimental to pond life. The suppliers of the low energy aerator provide instructions for building a simple, cheap filter that can be linked to their aerator. There are other purpose-

A garden pool filter.

built filters on the market, such as the Filtapac submersible pond filter, which is effective and easy to clean (see illustration).

First-aid when oxygen supplies fail

In the event of a prolonged breakdown, particularly during hot weather when fish might show signs of distress, it might be necessary to give first-aid. If the stoppage is caused by a power failure, spray the surface of the water from the nozzle of a garden hose held high above the water's surface. If the trouble is solely mechanical, a pump that operates on an electric hand drill can be used temporarily (see pages 19 and 20).

5 Water Plants

Aquatic plants play an important part in keeping the balance of pond life, and add interest and charm to a pool.

The large leaves of water lilies provide some shade for the fish in very hot and sunny weather; floating plants do the same thing. They provide food for fish such as carp, which need a considerable amount of vegetable material in their diet. On the other hand the plants feed and harbour a number of insects, which supply food for fish like rainbow trout that include them in their normal diet, especially when young. They give admirable spawning grounds and shelter for young fish. Marginal plants also provide areas of shade, and do not have the disadvantages associated with overhanging trees. Finally, plants in the process of photosynthesis yield oxygen, which is essential to the health of fish. There are certain water plants that yield extra-large quantities of oxygen and these are known as oxygenators. Among these are *Elodea canadensis*, *Elodea densa* and *Potamogeton crispa*—all varieties of pondweed. Sometimes oxygenators are supplied in bunches of mixed varieties. They grow completely submerged in the water. The recommended rate of planting is one bunch per 2 sq. ft (1,860 sq. cm) of water surface area.

Water plants in a fish pool assist in the control of algae, because they compete for the nitrogen and mineral salts that the algae need to thrive. The more common algae that worry many fishpond owners are blanket weed and a microscopic type that imparts a pea-green or sometimes copper-red tint to the water. Algae is not detrimental to plant or fish life, but many people find its presence an eyesore. Actually fish flourish on it, so it is best to leave it alone except for occasionally thinning out the more rampant growers, such as blanket weed.

Planting a pool—general advice

The dimensions of the pool are usually the criteria that determine what species and varieties can be grown. Readers are recommended to consult the catalogues of specialist water-garden nurserymen. Growers often provide collections, graded for varying sizes and depths of ponds and including an appropriate number of ramshorn snails which are invaluable for keeping the water clean and clear by eating all sorts of decaying matter. Once introduced they rarely need replenishing because they multiply very fast despite the fact that their eggs and young provide very nutritious food for fish.

A typical example of a collection for a pool with a water surface of 90 sq. ft (8.4 sq. m) and a depth of 18 ins (45 cm), about the same dimensions as the minimum size that is being considered here, is given below:

4 water lilies
48 oxygenating plants
12 groups of 2 each marginal plants
6 deep marginal plants
3 each of 4 floating plants
36 snails

Remember that although fish, and rainbow trout in particular, enjoy water in which there are gentle currents, most aquatic plants, including water lilies and water hawthorn, dislike stream effects, especially if the currents are cold. It is thus important to ensure that they are not placed in a direct current and to avoid introducing cold water from the mains except when it is absolutely necessary, such as during very hot weather or to prevent pollution.

Water plants are best grown in rich loam. If this is

not available, any good garden soil will do, particularly if it can be fairly substantially mixed with clay.

They are transplanted in the growing season, i.e. April, sometimes May, to early September, but it depends upon the weather. Spring frosts can delay planting by three to four weeks.

It is possible to place a layer of soil at the bottom of the pool, but this can cause the water to become murky if the fish disturb it. The greatest disadvantage, however, is that oxygenators dealt with in this way grow so vigorously that they soon completely fill the pond and choke the other aquatics.

It is usual nowadays to plant out in long-lasting plastic crates (see photograph), which are perforated on the sides and bottom to allow the roots free access to the water. They are designed with sloping sides which fit directly against the wall of the pool constructed as in Chapter 2. The crates can be placed on the bottom of the pool or raised on a platform of bricks to the depth required by the particular plant you are planting out. Fill the crates with soil to 1 in.

Planting crates of various sizes.

(2.5 cm) from the top. If garden soil is used, place hessian squares inside as a lining to contain it. When the crate is planted out put a layer of washed gravel over the top surface of the soil. This gives a finished appearance and prevents the fish from disturbing the soil and the roots.

Planting details for various groups of plants

Submerged oxygenating plants

These important plants can be quite simply planted in small plastic pots or baskets. If they are bought planted up they usually have a lead weight attached to them so that they sink to the bottom. If you plant them into your own pots in the garden, a weight such as a heavy stone should be tied to each pot. Alternatively they can be weighted and dropped unpotted on to a ½-in. (1.25-cm) layer of washed gravel spread over the bottom of the pool, in which they will root. This method prevents them becoming too invasive. As they have little root they can be planted at any time.

Marginal plants

These are planted in small crates or plastic pots positioned around the edge of the pool and supported on bricks to give them the correct depth of water, which usually varies from nothing to 6 ins (15 cm). Typical marginal plants are *Acorus calamus* 'Variegatus', depth 3–5 ins (7.5–12.5 cm); *Caltha palustris* (Marsh Marigold or Kingcup), depth from nothing up to 3 ins (7.5 cm); and *Iris laevigata* 'Snowdrift', 2–4 ins (5–10 cm). As a guide one plant should be provided for every 5 sq. ft (4650 sq. cm) of water surface.

Deep marginal plants

These are planted in polythene crates supported on

bricks a few inches below the surface of the water, and gradually lowered until they reach their required depth. Popular deep marginal plants are *Aponogeton distachyus* (Water Hawthorn), which requires a depth of about 12 ins (30 cm) and *Zantedeschia aethiopica* (Arum Lily), which must have its crown 6 ins (15 cm) below the surface of the water in order to survive frost.

Floating plants
Excellent representatives of this group are *Eichhornia crassipes* (Water Hyacinth), which is damaged by frost and should be wintered indoors. *Hydrocharis morsus-ranae* (Frog-bit) and *Stratiotes aloides* (Water Soldier or Water Cactus). These plants are merely placed on the surface of the water at any time during the growing season.

Water lilies
These beautiful plants are planted in crates supported on bricks so as to allow at first 3–6 ins (7.5–15 cm) of water above them. When they commence to grow they should be gradually lowered until they are at their right depth. When first planted, all the mature leaves and surplus roots should be removed with a sharp knife.

Maintenance of aquatic plants

Water lilies in containers must be repotted between mid-March and June every three or four years. Marginal plants should be thinned and replanted periodically.

In the autumn all excess growth and dead leaves etc. must be cut away from water plants with a sharp knife to prevent pollution.

Water plants growing in baskets and pots should be fed with fertilizer from time to time. This may be done by placing a ball, about 1½ ins (3.75 cm) in diameter, made of equal parts clay and bonemeal, into the soil alongside the plant roots. Alternatively you can buy a proprietary fertilizer in a perforated sachet, which is placed near the roots of the plants.

Pests

Pests that attack aquatic plants are water-lily blackfly (aphids), water-lily beetle and brown china marks moth. Fortunately these pests are consumed by the fish. Sometimes, however, it is necessary to hose them off plants so that the fish can get at them.

Wild aquatic plants

Do not introduce plants from rivers and streams into a garden pool, since plants from wild sources might bring diseases and parasites.

6 General Management of the Fishpond

Raising fish in a garden pond usually entails much less work than attending to a vegetable patch of much the same size. Certainly there are no essential daily or weekly chores, as with chickens and rabbits. The tasks that have to be carried out can be divided under three main headings. Firstly, it is necessary to look after the fish themselves, keeping them happy and healthy; secondly, attention must be paid to maintaining their environment in first-class condition, which means keeping the fabric of the pool clean and sound and the water clean and free from decaying organic matter; and thirdly, it is important to look after the water plants. Although there is some overlap it is convenient to divide the various tasks into seasons.

Spring

At this time of year there is relatively little work to be done. Divide and replant any aquatic plants that have made too much growth. To do this you must lift the crates out of the water.

Summer

The rather greater number of jobs that have to be carried out in summer are largely directed towards the well-being of the fish.

Loss by evaporation

When the weather is hot and sunny, evaporation can reduce the water level in a pool and it will be necessary to top up the water periodically. When first constructing a pool it is a good idea to lay a permanent plastic pipeline buried a few inches below the ground and connected to a standpipe for this purpose and also for cooling the water. If this has not been done, get out the garden hose and replenish the water when the level falls. In hot weather, the loss by evaporation can lower the surface of the water by 2 ins (5 cm) in a week. If it falls at a greater rate than this, suspect a leak. This can be dealt with later in the year, unless it is very serious. Replacing water is particularly important when the pool has a plastic liner, because exposure to sunlight perishes it and causes it to become brittle.

Shortage of oxygen in the water

Another problem that can be caused by hot summer weather is a shortage of dissolved oxygen in the water, because gases are less soluble in warm water than in cold and are consequently driven off as the temperature rises. This is contrary to what happens in air, in which the proportion of oxygen remains constant whatever the temperature. It is important, therefore, to keep fountains, aerators and waterfall units running for as long as possible, so that they can restore the oxygen balance. This is particularly important at night, when the oxygen content of the water tends to fall. If there is a shortage of oxygen in the water (either because the various aerators are not adequate or because there is a temporary breakdown) the fish will come to the surface of the water, gasping and gulping the air. If this happens, spray the surface with a garden hose from a height of at least 3 ft (0.9 m). The higher the pressure of the jet the better. If only a single fish is behaving in this manner, the oxygen content of the water is probably satisfactory and the particular fish may be unwell.

Pests on water plants

During the summer certain pests such as blackfly (aphids) and water-lily beetle will infest aquatic plants. Keep a sharp look-out for them and wash them into the water with a strong jet from a garden hose; the fish will gobble them up. Timely action of this sort will also prevent damaged plant tissue becoming rotten and turning into a source of water pollution.

Removal of algae

A form of algae called blanket weed, which is a filamentous type, can be a nuisance in a pool if it grows too much in sunny weather. Its presence is not detrimental to plants or fish, but an abundance of it can look unsightly. Its growth can be minimized, except in very small pools, by growing other water plants. It is possible to remove it mechanically by simply poking a bamboo cane with a split end into the mass and slowly winding the filaments on to it. There are several proprietary preparations which can be added to the water to eliminate blanket weed, but before doing so it is recommended that as much as possible should be removed mechanically. Take care to see that whatever is used is not harmful to fish or plants. When using any such preparation it is important to know the cubic capacity of the pool so that the correct dosage is put into the water. Too much could be harmful to both the fish and the plants.

Finally when a chemical is used for destroying algae it is important to remove the filament masses as soon as they turn brown to prevent them blanketing other plants and also to avoid oxygen deficiency in the water.

Autumn and winter

Autumn and winter are undoubtedly the periods when there is most to do in the maintenance of a garden fish pool. General repairs (see page 21) can be carried out at this time and the more specific tasks are listed below:

Pollution prevention

The first important thing that has to be done to maintain a healthy colony of fish is to see that pollution does not affect the condition of the water, because the toxic substances from it are the commonest cause of fish dying. Water is usually polluted by decaying vegetable matter, such as fallen leaves, branches and twigs. In a garden pool these come from two main sources—dying water plants, especially the large lush leaves of water lilies, and the leaves of deciduous trees and shrubs; and, perhaps to a lesser extent these days, soot from the atmosphere. Other sources of danger that must be closely guarded against are dead frogs and fish.

As mentioned previously two of these sources of trouble can be averted by anticipation. Always ensure that your pond filter is free from debris and from August onwards look for dying leaves, stalks, and seedpods on water lilies; deep marginal water plants and other aquatics are likely to discard them generously. As they appear they should be removed. When the leaves of marginal plants die off, usually in October, they should be cut off.

Leaves from neighbouring trees and shrubs, the other serious source of pollution in a fishpond, should be prevented from falling into the water by spreading a net over it in good time. Such a net might also prevent the fish being stolen by herons, kingfishers and other predators. Typical nets range in size from 12 × 9 ft (3.6 × 2.7 m) to 36 × 12 ft (10.8 × 3.6 m). They are made of ¾-in. (1.9-cm) plastic mesh. Spread the net right over the pool. If the surrounds are soft ground, fix it firmly in position by pegs supplied with the net. Otherwise hold it down with bricks or pieces of stone. Normally it is placed in position after the pool heater (if you are using one) has been installed.

If leaves have already fallen into the pool, possibly because of an unexpected gale, before the net has been put into position, they can be removed by dredging the bottom with a fish net. This can be done more effectively if the pool is first half-emptied. Sometimes leaves and debris collect on the surface of the water before they sink to the bottom. These can be removed with a hand fish net. An improvised device for this purpose consists of a piece of chicken wire fixed over the prongs of a garden fork. The netting should be about 16 ins (40 cm) wider than the prongs and should have the last 5 ins (12.5 cm) on either side bent up at an angle of about 20° to the vertical. Take great care when using this device in a plastic-lined pool.

Regular partial replenishment of water with fresh supplies also assists in keeping down the concentration of toxic substances. In fact, if the water looks dark green or blackish in October and you do not want to empty the pool completely, the water can be pumped out to the half-way mark and replaced with fresh.

Prevention of freezing

Another essential task for late autumn and winter is to do everything possible to keep the pool from freezing over. While a layer of ice does not directly hurt the fish, it seals in toxic gases arising from decaying organic matter, and carbon dioxide, which the fish continuously breathe out and which is produced by water plants at night. Also an ice layer would prevent supplies of oxygen entering the water from the atmosphere, to the great detriment of the fish.

Furthermore, while it does not affect the waterproofing of a pool lined with butyl synthetic rubber or PVC, the expansion of the water on freezing might crack a concrete one and cause it to leak. If there is no other means of preventing this, a tennis ball floating on the surface of the water will often avert this kind of damage.

In water gardens which contain waterfalls and fountains, freezing can be avoided, unless the frost is very severe indeed, by running the pump continuously to keep the water on the move. This also applies to pools in which a modern aerator has been installed. Since a garden pool is usually in an open, sunny position, this action, together with the warmth of the winter sun, is often enough to keep the pool from freezing over.

As there is usually a supply of electricity available at a pool, it is a wise precaution to install pool heaters. This apparatus is not intended to heat all the water, but to keep a small area of the surface free from ice so that any toxic gases and the exhalation of fish and plants are able to escape. It also allows some of the surface to be exposed to the air, and so oxygen can be absorbed. The makers recommend that you install one pool heater for every 30 sq. ft (2.8 sq. m) of surface area. One might usefully be placed close to the inflow and outflow of the pump circuit so that there can be a continuous flow of water.

Pool heaters are purpose-designed and consist of a nickel-chrome element wound on a sealed ceramic former encased in a nickel-brass tube (see Fig. 15). They have a loading of 125 watts. The container is fixed in a block of expanded polystyrene, which allows the heater to float and holds it in a vertical position. Sealed into the heater is a 6-ft (1.8-m) length of cable. This is long enough to reach a convenient place outside the pool where the heater can be joined to a source of mains power by means of a weatherproof connector.

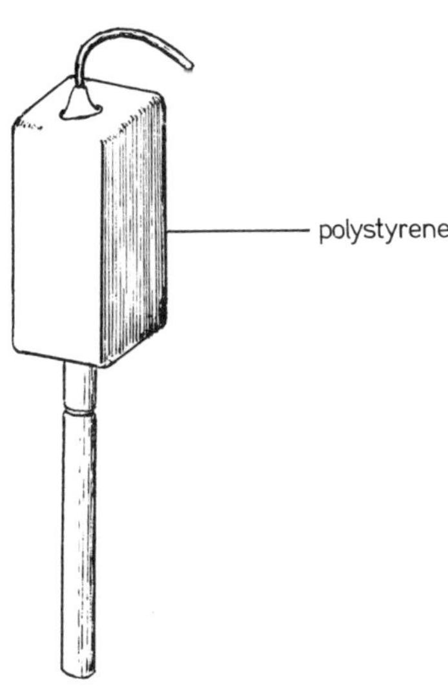

Fig. 15 Pool heater.

The heater should be put in position before any severe frost occurs, and before the pool is covered with a net to collect falling leaves.

Never break a thick layer of surface ice with a heavy instrument. Apart from the risk of injuring the fish, mixing ice and water lowers the temperature of the water too much. In addition, the valuable insulating power of this ice layer is lost.

If you need to make a hole in the ice, stand a metal container on the ice and fill it repeatedly with boiling water. After a time the ice will melt, forming a hole in the surface.

Renovation of pool liner

After removing any fish (see page 79), repairs can be carried out as described in Chapter 2.

Overhauling mechanical equipment

Late autumn is a good time to check the pumps, clean

them and have them overhauled. At the same time, examine fountain jets and clean out any holes that have become stopped up. Also check all the electrical cables and connections. Give this equipment very special attention: because of the proximity of the water in the pool it would be dangerous if there were a breakdown in the insulation. It might be advisable to ask an electrician to check the circuits.

Late November to early December is a very suitable time to examine the equipment of a fish pool, because the low temperature at this time of year means that oxygen concentration in the water is relatively high. Also fish at this time are rather less lively and their demand for oxygen is less. This makes it easier to shut off the pumping plant for a short time. Nevertheless the fish must be kept under observation. If any signs of oxygen starvation appear, deal with the situation immediately by playing a jet of water on to the surface from the nozzle of a garden hose held about 3 ft (0.9 m) above the water.

7 How Fish Live

A fish is a cold-blooded creature which breathes in oxygen extracted from the water, not from the air as in the case of mammals. There are three classes of fish and they are primarily differentiated by the type of skeleton that they have. Most British fresh-water fish (including rainbow trout, carp and tench) fall into the third group, the members of which have jaws and bony skeletons, while those of the other classes have cartilage skeletons, with jaws in some cases and without in others. Many of these fresh-water fish with skeletons of bone structure have swim bladders, the function of which is discussed overleaf.

Body shape of fish

There are great variations in the body shape of fish over the whole range of species. The most usual, however, is a streamlined one like that of the typical fish (shown in Fig. 16). Certainly this is the approximate shape of rainbow trout, carp and tench.

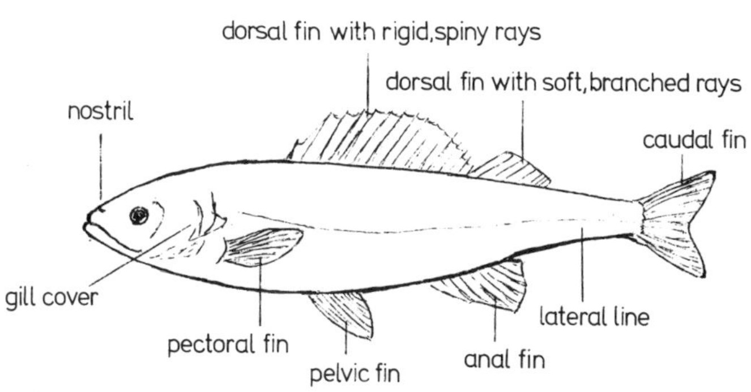

Fig. 16. Typical fish.

How fish move

As mentioned before, some fish have a swim bladder, which is situated in the middle of their body. It is a silvery, gas-filled bag which functions like a buoyancy tank in a submarine, permitting a fish to float and swim at any depth in the water without rising or sinking. Any fish that have not got this organ must keep on the move continuously.

A fish can follow any course it wants through the water by using its fins. The majority of fish have two pectoral fins, two pelvic and one each of dorsal, caudal (or tail) and anal fins. This, however, varies from species to species and some fish lack pelvic or pectoral fins, while others lack a dorsal or caudal fin. In some instances several fins are amalgamated. An example of this is in the eel, whose very long dorsal and anal fins merge with the caudal fin, giving a continuous edging of fin along the whole body. Using its fins a fish moves and swims through the water by means of wriggling movements of its body. The fins control stability and direction.

The senses

Fish have eyes with a very wide range of vision. Between the eyes and the mouth, on the uppermost side of the head, are two small vents—nostrils, but the fish does not breathe through these. Instead they are used for smelling only and have nothing to do with the respiratory system.

There is a faint line running on either side of the body of a fish, which is known as the lateral line. This consists of a series of highly sensitive cells, which, by means of records of pressure, indicate distances and warn of hazards. It is, in fact, a built-in radar system. In addition, some fish have barbels, which are fleshy filaments hanging from their mouths, which they use as feelers.

Breathing

Fish have a special mechanism in the shape of their gills for extracting oxygen from the water. The gills are positioned on either side of the head, a little behind the eyes (see Fig. 16). The vital parts of the gill, the gill filament and the gill lamella, are protected by the gill cover. The water enters the gill by way of the mouth, which contains what amounts to a non-return valve preventing the water from flowing out. It leaves through the skin fold, which functions as a valve at the opening at the rear of the gill cover. Any food particles are sieved out by the gill rakers on the inside of the gill arch. The water then flows through the gill lamella, in which there is a fine network of blood capillaries to absorb the oxygen in the water into the bloodstream of the fish.

Feeding

Many fish are carnivorous, but they do not all prey on other fish. The most usual food consists of a variety of bottom-living and mid-water organisms, such as insect larvae, molluscs, small planktonic crustaceans, etc. Some fish live on insects which fall on to the water surface. Often, when these are plentiful, they do not touch other food. Some species of fresh-water fish live exclusively on vegetation. Carp and tench like to eat water plants.

The digestive system

As a fish does not breathe through its nostrils or its mouth, but by means of its gills, its gullet or oesophagus leads from the mouth straight into the body cavity (see Fig. 17). Usually a fish does not chew its food, its teeth being used to grab food or prey. The food is swallowed whole. Consequently the gullet is capable of stretching to a considerable extent. Food passes next into the stomach, which is folded

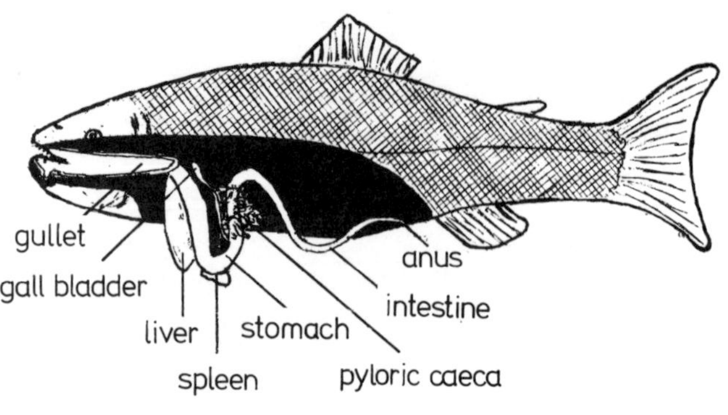

Fig. 17 Digestive system of a fish.

into a U-shape. The first half (cardiac), which is the larger, is capable of being distended considerably by a good meal. The second half (pyloric) leads through a sphincter (circular) muscle, which joins the stomach and the intestine. Just beyond is the pyloric caeca, which has numerous long, narrow sacs in it, rather like the fingers of a rubber glove. There are usually more than thirty in a rainbow trout. These increase the surface of the absorptive area of the gut and sometimes aid digestion by secreting digestive juices. The length of the intestine varies appreciably from one fish to another. It exists in the cavity of the body of the fish as a loosely coiled tube. At its far end it opens to the outside by way of the anus, which is situated just behind the pelvic fin.

At the forward end of the body cavity is the bi-lobed liver with the gall bladder from which the bile duct leads to the intestine. In this same area there are scattered the pancreatic tissues. The spleen is joined to the back arm of the stomach.

The digestive juices are secreted into the stomach, the pyloric caeca and the front end of the intestine.

Other secretions are injected from the pancreas area. Bile, which emulsifies the oils in the food, is injected from the liver through the bile duct. These all contrive to break down the various foods into compounds that can be assimilated by the body of the fish through cells in the walls of the intestine. The unwanted, undigested remains pass out as faeces through the anus.

Environment

Temperature of the water

The temperature of the water in which fish live is very important. In the first place it has a very considerable influence on its oxygen content, because the solubility of any gas decreases as the temperature rises. Water that is satisfactory for a particular species when it is colder might prove lethal when it becomes too hot because the oxygen has been driven off.

The optimum temperature of the water varies for different fish, so they will not naturally live together in the same environment. As an example, that for brown trout is between 53°F (12°C) and 60°F (15°C), whereas that for carp is between 72°F (22°C) and 77°F (25°C). These figures are consistent with the fact that brown trout revel in the cold, well-oxygenated waters of mountain streams, while carp are happy in warm, lowland ponds in which oxygen is much less abundant.

Rainbow trout tolerate higher temperatures and lower oxygen concentrations than brown trout. Consequently it is possible to come to a compromise and have some ponds in which rainbow trout and carp will live reasonably well together and flourish.

The temperature of the water has a most important bearing on the supply of fish food. Natural fish food is much scarcer in cold years than in warm ones. Thus water temperature must be taken into account when keeping fish domestically. When the weather is warm there are ample supplies of food in the pond and it is not necessary to supplement them with domestic scraps and purchased foods. If these are given, the fish are likely to leave them, and the surplus food will become putrid and pollute the water.

Finally (although it may not be of such great consequence when carp and rainbow trout are farmed, because they do not breed freely in Britain), unless the proper conditions are provided, as at commercial hatcheries, the temperature of the water can considerably influence the various natural processes of breeding. It must be at the appropriate level to start the reproduction cycle. It also determines the length of time the fish embryos take to develop from the egg. Probably the reason why fish like carp and trout do not breed easily in ponds is that they need a different temperature at spawning time than is normally available in the open in Britain. It is known, for example, that carp will only spawn in water whose temperature rises to more than 57°F (14°C).

Fresh, clean water

The water in a pool containing fresh-water fish must be fresh. On no account must it become brackish (there is a rare possibility of this happening if the pool is replenished from an old well). It must not be allowed to become stagnant, because it might contain substances obnoxious to fish. There must be no decaying organic matter present as this produces toxic gases harmful to fish.

8 Fish Suitable for Garden Fish Farming

There are three fish that are regarded as suitable for raising in a garden pool—these are rainbow trout, carp and tench. Perhaps the most attractive is the rainbow trout, on account of its pleasant colouring and because it creates considerable interest by rising to the surface of the water to catch insects. The tench, on the other hand, only moves about at night; during the day it spends all its time in the mud or among the weeds at the bottom of the pool and is seldom seen. If it is tempted to approach the surface of the water on a calm summer day and rest under the flags or water lilies, as soon as it is disturbed it darts to the bottom of the pool and buries itself in the mud. Between these two falls the carp, which although it likes ponds with a muddy bottom in which it finds some of its food, will also rise to the surface and bask in the sun. It will accept small pieces of bread or cake that are thrown to it.

Rainbow trout

This fish was introduced into British waters from western parts of the USA at the end of the nineteenth century. It is an attractive fish, with a lateral pink or red stripe that runs along its silver sides. Its sides, back, dorsal and caudal fins are densely covered with black spots. To some extent it resembles the British brown trout in appearance and habit, but it is distinctly more colourful. Other important differences are that the rainbow trout will tolerate a rather higher temperature, a rather lower oxygen content in the water and somewhat less clear water than the brown trout, which revels in cold, sparkling waters. These differences make it possible to keep rainbow trout in comparatively small garden ponds, in which the small volume of water gives rise to wider temperature fluctuations. In occasional heat-wave conditions, when the temperature might rise to a dangerously high level, it is wise to have facilities available to cool the water by removing some of it and slowly feeding in cooler water from the mains. It also helps to shade the pool with hessian screens on very hot days.

Rainbow trout feed naturally on a range of adult insects and larvae—the mayfly is a particular favourite. The older ones prey on other fish. When they are kept in a pool in a garden they will find a considerable part of their diet for themselves, but often, in order to obtain a good growth rate, it is necessary to supplement this. They can be given household scraps—especially if the scraps contain a small amount of meat; chopped-up earthworms, which are a particular favourite; maggots, which can usually be obtained from a fishing-tackle shop; minced meat—the cheapest kinds, of course, such as lights; fish-meal pellets and pellets containing a complete ration.

Like other fish, given an unlimited supply trout always eat a definite amount, which might differ from day to day, but which is relatively constant over a longer period (such as a week or a month). Research, however, suggests that if the amount of food that is allowed is limited so that their appetite is not quite fully satisfied they automatically use the food more efficiently. So do not over-feed them, because they grow equally well when the quantity of food is restricted. If possible, try to observe how much they will eat voluntarily, and then cut the amount given to just a little below this.

Rainbow trout grow fast. They will reach a weight of over 2 lb (1 kg) in their third year. Within three or four months a 3–4 oz (84–112 g) trout can weigh over ½ lb (227 g). They are relatively short-lived, with a life-span of about four to five years. Their biggest disadvantage is their instinct to migrate to the sea. This is not, however, likely to give any trouble in most garden pools, because they are land-locked and the fish cannot escape. Nevertheless it might be a problem in a fish pond with a stream running through it. Some populations, however, are non-migratory, so discuss the matter with your supplier before you stock your pool.

Carp

In their natural environment carp will prosper in small, muddy, stagnant pools, where they tolerate fairly high temperatures and low oxygen supplies compared with the needs of rainbow trout. So if for any reason you cannot oxygenate your garden pool mechanically, carp is a particularly good choice, providing the pond contains ample vegetation, including oxygenating plants.

There are two forms of carp that have become particularly domesticated—these are the leather carp and the mirror carp. Both are forms of the fast-growing king carp. The leather carp is practically devoid of scales, while the mirror carp has much enlarged scales confined to one or two rows on each side of the body. They are fast-growing, long-lived fish. They grow 1–3 lb (0.5–1.5 kg) in weight per year and weigh an average of 2½ lb (1.1 kg) at the end of the third year. As a guide, a 12-in. (30-cm) carp weighs a little over 1 lb (454 g), while a 15-in. (37.5-cm) fish turns the scales at about 2¼ lb (1 kg).

Carp prefer a pond with a muddy bottom. So unless

a garden pool is a natural one, it is an advantage to put a 3–4-in. (7.5–10-cm) layer of soil over the bottom.

The carp's natural diet is largely vegetable, but it does include a proportion of worms, insects and their larvae. They obtain these largely from the mud, which they even take in and ultimately reject. Thus it is important to have plenty of aquatic plants in a pool containing carp.

The natural rations found by the carp should be supplemented with household scraps. Carp can cost nothing at all to feed if there are enough scraps. Otherwise they can be given boiled potatoes, flaked maize, mashed peas, grain, ground barley and proprietary fish food. Once again, they must not be overfed.

Tench

Tench live in the same conditions as carp. They are probably the most sluggish of fresh-water fish, so much so that they can be easily caught by hand if you grope cautiously in the mud. As with carp, a layer of mud should be put on the bottom of an artificially-made pool intended for tench.

It is a sturdy fish with unusually small eyes and a small barbel (feeler) at each corner of its mouth. Its back is brownish-green, its sides have a golden sheen and its belly is creamy-yellow. Its scales are very small and numerous. A 12-in. (30-cm) tench weighs a little more than 1 lb (454 g) and a 15-in. (37.5 cm) one about 2 lb (1 kg).

When fending for itself, the tench eats plants, worms, insects, larvae and snails. When kept in a garden pool, which should be amply stocked with aquatic plants, tench should be given supplementary food similar to that given to carp. Remember that in cold and very warm periods it stops eating and buries itself in the mud.

Stocking a fish pool

In a newly constructed fishpond, fish may be introduced about two weeks after planting, by which time the oxygenating plants should be established. In an old pool with existing plants the fish can be put in immediately. When introducing fish into a new pool, you must provide supplementary food for about a year or until surplus food is left by the fish, whichever is the sooner. Doing this will help insects etc. to reproduce to provide natural food.

As a rough guide, introduce one fish per 2 sq. ft (1860 sq. cm) irrespective of the size of the fish when they are first introduced. There will then be ample space for growth.

The size of the new stock can vary quite considerably. Naturally the smaller ones are cheaper and the larger ones more expensive, but the latter are ready for the table earlier. Suggested sizes are rainbow trout 2½–3 ins (6.25–7.5 cm) long; carp 2½–4 ins (6.25–10 cm) long; and tench 2½–4 ins (6.25–10 cm) long. Be guided on the question of size by the supplier, who is familiar with his stock.

Small carp and tench of the sizes given above can be safely introduced between September and April. Some people prefer to introduce them later rather than earlier, possibly in April. Rainbow trout are also better moved at this later date. This, however, is a matter on which it is best to seek the advice of the supplier when ordering.

When a sizable quantity is involved, fish are delivered in well-oxygenated tanks, by road. Very small orders of young fish are packed in oxygen-filled polythene bags inside strong cardboard boxes, and despatched by express passenger train to the buyer's nearest railway station for collection. The buyer is informed by telephone beforehand. Delivery of fish is very expensive and it is better wherever possible to collect them from the nearest fish farm by prior arrangement.

When the fish arrive they should be put into the pool as soon as possible. They should not be handled because this is likely to injure them. The container in which they are delivered should be opened and placed under water. It is possible to put some carp into the same pool as rainbow trout, providing the pool is not allowed to get very warm and is kept well oxygenated. It is suggested that the percentage of carp should be not more than 40 per cent of the total.

Where to buy fish

Although numbers of garden centres, commercial water gardens and pet stores stock ornamental fish, very few sell edible fresh-water fish. Some, however, have contacts and can obtain supplies for a customer.

Throughout the country there are numerous fish farms from which trout, carp and tench can be bought. Numbers of these advertise in the sporting magazines. It is cheapest to find a local supplier from whom you can collect, so use the telephone directory's Yellow Pages. Although it is not exhaustive, a list of names and addresses of selected fish farms throughout Britain is given in the Appendix.

9 Looking After Fish

Feeding fish

In well-established garden ponds, fish can survive without being fed from the outside if there is plenty of vegetation on which water insects can breed. The one exception is when fish are introduced into a newly-made pool, because the vegetation and other natural foods have not built up to the extent needed to sustain fish without some additional contribution. Supplementary feeding should be carried on for at least the first season after stocking such a pond.

The food which various types of fish eat differs to some extent. Several, such as perch, pike and zander, live largely on other fish, even to the extent of cannibalism of their own species. It is important therefore that such predators are not introduced into a garden pool stocked with, say, carp or rainbow trout, which, although they will eat fry, largely live on insects, larvae etc. and, especially with carp, pond vegetation.

There are some fairly well-defined rules covering the feeding of fish kept in a garden pool, whether for decorative or edible purposes. To produce healthy fish these rules must be adhered to strictly.

In the first place, if fish are to grow quickly for the table, it is essential that they should have ample oxygen. Another factor is the temperature of the water. When it is warm their needs are greater than when it cools down in winter. In fact when it is really cold their food intake falls almost to nothing. Between late November and early March it is seldom necessary to feed fish at all. There are, however, two times in the year when it is important to be sure that they are well fed. One of these is at the end of summer and early autumn—September and October—when a supplementary high-protein ration should be provided, because natural life in ponds diminishes and it is important for the fish to be fortified to withstand the winter, during which time they rely on nourishment stored in their own bodies. The other time is in the spring, when insects and other natural foods are only just beginning to come to life again. At these two periods it is an advantage to supplement their natural diet with chopped-up earthworms, vegetable proteins, yeast products, shredded meat etc. There is an excellent proprietary high-protein food in the form of floating pellets, produced by a leading manufacturer of trout and salmon food. It is a complete diet of fish meal, vegetable proteins, yeast and cereals to which is added a proportion of shrimp meal, vitamins and mineral. As the pellets float, the fish come to the surface to nibble them. The pellets are easily removed if they are not soon devoured, so there is no risk of pollution. During the rest of the year other food that can be given includes baked breadcrumbs, oatmeal soaked in hot water, mashed peas, cooked potatoes and, cheapest of all, household scraps.

If fish are being raised for the table and are to grow satisfactorily in a reasonably short time, they must be given an ample amount of food in addition to that which they get from pond life. The operative word is 'ample'. As a guide, they should have no more food than they will consume in five minutes. The amount fish will eat in this period depends on a number of factors, such as whether they are keen and active, and the temperature of the water. Normally, one feed a day, if it is of the right size, is sufficient.

There is one danger period—when you go away for two or three weeks' holiday. A kindly offer from a neighbour to feed the fish should be gently refused, because the fish may well be inadvertently overfed. It is quite safe to leave the fish to fend for themselves.

Feedometer

While it is probably better to regulate the rate of feeding fish by experience, there is one device that can give some guidance—a Feedometer. This is a temperature gauge which indicates when and when not to feed the fish. (See illustration.)

Fish feeder

You can also buy a floating food dispenser, which distributes food automatically when it is activated by the fish. It is suitable for all types of feeding stuff and for all sizes of fish. They take just what they want and there is no risk of pollution. This device is excellent for use during holiday times and also during mild periods in winter. In a larger pool it might be necessary to install several of these floating fish feeders.

Removing fish from a pool

If ever you have to empty a pond containing fish you must remove them to a safe place such as a large barrel

A feedometer.

or water butt, or a temporary pond constructed in a corner of the garden by digging a hole of a suitable size and lining it with polythene sheeting. If for any reason the fish have to stay out of the main pool for a long time, a temporary pond is the best solution. It is important, however, to make sure that the temporary pond is large enough, so there is no danger of overcrowding.

When removing fish from the pool never catch them in your hand as this might damage their scales and remove the protective mucus covering their bodies. This renders them more liable to disease. Fish must always be caught in a net. There are several designs of nets. Usually the framework and handle are made of aluminium. Satisfactory nets have an opening of 120 sq. ins (770 sq. cm). There are also nets available with 7½-ft (2.25-m) long handles, which are telescopic. In addition there are hand fish nets which have a wire framework and a rot-proof, very fine mesh. Hand nets vary in size, but the largest opening is about 10 × 9 ins (25 × 22.5 cm). They are particularly useful for removing fry from the main pool to a nursery pond if breeding takes place. If the fry are not taken out, until they reach more than 1 in. (2.5 cm) long they are likely to be devoured by the larger fish.

0 Predators, Pests and Diseases

Predators

Herons

Herons are probably responsible for a large proportion of fish taken from garden pools. They stand up to 3 ft (0.9 m) high. They usually nest at the top of tall trees, particularly oaks and elms. If there are any of these trees in the vicinity and herons have attacked fish in a pool at any time before, they are likely to do so for some time, because they return to the same heronry each year. Another particularly vulnerable area is one in which there are sea cliffs or reed beds, which are alternative nesting places for them. Their method of taking fish is to stand on one leg on the edge of a pool, with eyes half closed and head hunched between the shoulders. As soon as a fish comes within reach they will pounce on the victim, hitting it on the head with their long bill and stunning it so that they can carry the unconscious fish away.

The best protection to fish against herons is to keep the pool covered with a net in the same way as is done for keeping out falling leaves. The net should be tightly spread out at a height of about 6 ins (15 cm) above the maximum level of the water, so that the bill of a heron, which is several inches long, cannot reach the fish. Since herons are about all the year round, the pool must be covered permanently to secure full protection. A net so positioned will not seriously interfere with marginal plants, because after a short time they will grow through it. Alternatively, it is possible to purchase specially designed nets which cover the surface of the water for about 2 ft (60 cm) around the pool. These are effective because a heron usually stands on the edge of the pool. They are also rather less unsightly.

Kingfishers

The beautiful kingfisher, which nests along rivers and streams, has a large, dagger-shaped bill, which it uses to grab fish as it dives under the water in search of food. It usually conducts its onslaught from a nearby perch or by hovering over the water. Once a fish has been caught it returns to its perch, where it manipulates the body of its victim so that the head faces outwards. It kills or stuns the fish by banging its head against the perch or other hard surface, and then reverses it and swallows it head first. Once again a secure net affords good protection, but the surface must be fully covered.

Cats

Domestic cats are sometimes attracted towards live fish in a garden fish pool. Here again a net is a good protection, particularly because it is very unstable for the cat to walk on. There are a number of proprietary cat repellents and specially prepared pepper dusts available nowadays that are quite effective in keeping cats away.

Mink

Another possible danger to fish is the mink, which is mainly an aquatic animal and, in the wild, catches most of its food in the water. Any that are free have escaped from mink farms. They can be prevented from entering a garden pool by a securely fitted net. Their risk to domestic livestock is recognized by the fact that there are stringent laws concerning the keeping of mink in captivity, and any occupier of land who knows that mink are at large on his property must notify the Ministry of Agriculture, Fisheries and

Food, who will arrange for the animals' destruction.

Great diving beetles (Dytiscus marginalis)
Many water beetles that inhabit a pond are harmless and provide fish food, but the great diving beetle is a dangerous creature. It is easily recognized by its blunt, oval, 1–1½ in. (2.5–3.75 cm) long and ½–¾ in. (1.2–2 cm) wide body. It is brown, with gold edging. Both the larvae and adults are carnivorous and will, if there is no more easily obtained food available, devour fairly large fish. They usually appear in May and June. Fortunately, as they have to rise to the surface for air, they can be removed with a net.

Other menaces to pond life

Frogs
Frogs are more a nuisance than a danger to fish, except that on rare occasions fish are injured by colliding with them. If, however, they are allowed to breed they can create a disturbance with their croaking. They are prolific breeders and a pool can soon become crowded with tadpoles. These should be removed with a net. Frogs live most of the year near water, but in February congregate in ponds to breed. The best way of dealing with frogs is to cover the pool with a net, particularly during late winter and spring.

Toads
Similarly, toads in a garden pool are more a nuisance than a menace, by filling it with tadpoles and by devouring food that is needed for the fish. They live mainly on dry land, but always return to the water in which they were born in late March and April for breeding. They can be kept out by covering the pond with a net.

Newts
Newts can be a danger to fish because they include

small fish in their diet. They live most of the year on dry land but enter the water to breed in early spring and leave it again at the end of the breeding season. Although they are usually rather smaller than mature frogs and toads a net will stop them entering a fish pool.

Great pond snail

Although most water snails do no harm, the great pond snail or fresh-water whelk should be removed from a pool if it appears because it does considerable damage to water plants. The snails can be recognized because they are longer (up to $2\frac{1}{2}$ ins/6.25 cm) than other water snails and they have a long narrow shell with a very pointed end. If any eggs appear on the water plants they should be hand-picked. The adult snails can be trapped by putting a lettuce or cabbage stump in the pool. This should be removed very frequently and any snails on it shaken off.

Parasites

Fish lice

These are tiny parasites $\frac{1}{8}$ in. (3 mm) long. They attach themselves to fish and feed from them. Affected fish often rub themselves violently against fixed objects. There are some proprietary remedies for treating them, but lice can be removed with tweezers from the body of an ailing fish, after touching them with a spot of paraffin. Vaseline should be smeared over the affected area.

Anchor worm

This is a white, worm-like parasite, $\frac{1}{4}$ in. (6 mm) long, which has an anchor-like head by which it attaches itself to its host. They should be removed like fish lice. Take care to see that the anchor head is extracted.

White spot

Although this condition is sometimes regarded as a disease, because of its appearance, it is really due to a parasite. A mass of tiny white pinheads appear all over the fins and body of the fish. These spots are caused by a protozoon, which is a primitive, one-celled, microscopic animal. These minute creatures attack the fish by burrowing into the skin, causing a sore about 1 mm in diameter to be formed. The parasite leaves the fish in about a week and forms a hard coating known as a cyst. Inside, the cyst divides into up to a thousand 'swarmers', which hatch after a few hours, and can attack the same or other fish. Sometimes the sores become infected with fungus and the fish can die. If any fish shows signs of attack, it should be taken out of the pool and kept in a clean tank of clean water changed every 12 hours, for at least a week.

There is a greater risk of disease spreading in a garden pool if an infection enters it, than in the wild. This is because diseased fish are weaker than healthy fish and are more easily caught in natural streams and lakes by predators, which are absent from garden pools. Thus only the fittest fish survive. It is important, therefore, to remove any lethargic or seedy fish immediately they are spotted, either for treatment or destruction.

Diseases

Fungus disease (cotton wool disease)

Fungus spores are present in all still waters. Fortunately, healthy fish have a film of mucus over the whole of their body, which keeps them healthy and protected from disease. If for any reason the fish is unhealthy and the supply of mucus is not being produced or the fish has been scarred by parasitical attack or damaged in any way, possibly by rough handling, the way is open for fungal infection to enter.

If there is any sign of a fish having this disease, it should be removed from the pond and treated. It can be placed in a bath of a proprietary fungus cure, made up in accordance with the manufacturer's directions. There are, however, other ways of treating this disease. The first is to place the affected fish in a salt bath, changed daily in strength:

1. First day—1 oz (28 g) of salt (rock, sea or cooking salt, *not* table salt, which contains magnesium carbonate) per 1 gallon (4.55 litres) of water.
2. Second day—2 oz (56 g) of salt per gallon of water.
3. Third day—3 oz (84 g) of salt per gallon of water.

The strength should be kept at this last level, changing the solution every day, until a cure has been effected. Then the fish can be returned to the pool. If it shows any sign of distress while it is in the salt bath it should be returned to fresh water without delay.

The second treatment, which must be carried out at the early stages of the disease, is to put the fish for half an hour in a solution of 1 g of potassium permanganate in 17½ pints (10 litres) of water. After this it should be put into fresh water and when it appears to be cured it should be returned to the pool. If it is practicable, if this disease appears, it is wise to change the water in the pool.

Tail rot

This disease affects the tail of the fish, giving it the appearance of being eaten away. It starts at the edge of the fin and gradually works into the body. To be effective early treatment is imperative. You can buy a proprietary preparation for dealing with it, otherwise any affected fish should be given a salt bath as described for fungus disease.

Dropsy

Little is known about this disease of fish and there is no satisfactory cure for it. The eyes of the fish protrude, the body becomes bloated and the scales stand out. Fortunately it occurs only occasionally, but if it does strike there is no alternative but to destroy all the affected fish.

11 Cooking Fresh-water Fish

Catching and killing

First you must catch your fish. This operation is best done by using one of the fish nets described on page 80. After catching, the fish will have to be killed. The preferred method is to deliver a sharp blow to the head of the fish, just behind the gills, using a blunt, heavy instrument such as a 'priest' or cosh.

Cleaning fish

Before any fish can be cooked it is most important that it is scaled, gutted and washed carefully. Gutting is quite simple and can be performed as follows:

1. Having removed the scales, lay the fish out on a flat surface such as a fairly large chopping board. First remove the head by cutting it off just behind the gill opening with a sharp knife (Fig. 18a).
2. Slit along the gullet and remove the intestine and the other internal organs from the body cavity.
3. Wash the fish out with cold water.
4. Remove the bones from trout, carp and tench, unless you want to cook them whole. There are two methods of doing this:

Method 1 (Fig. 18b–d)

(a) Split the fish along the belly. Open it out and lay it flat on a wooden or Formica surface with the inside downwards.
(b) Run your fingers along the backbone.
(c) Turn the fish over and it will be found that the bone and the ribs can be removed quite easily.

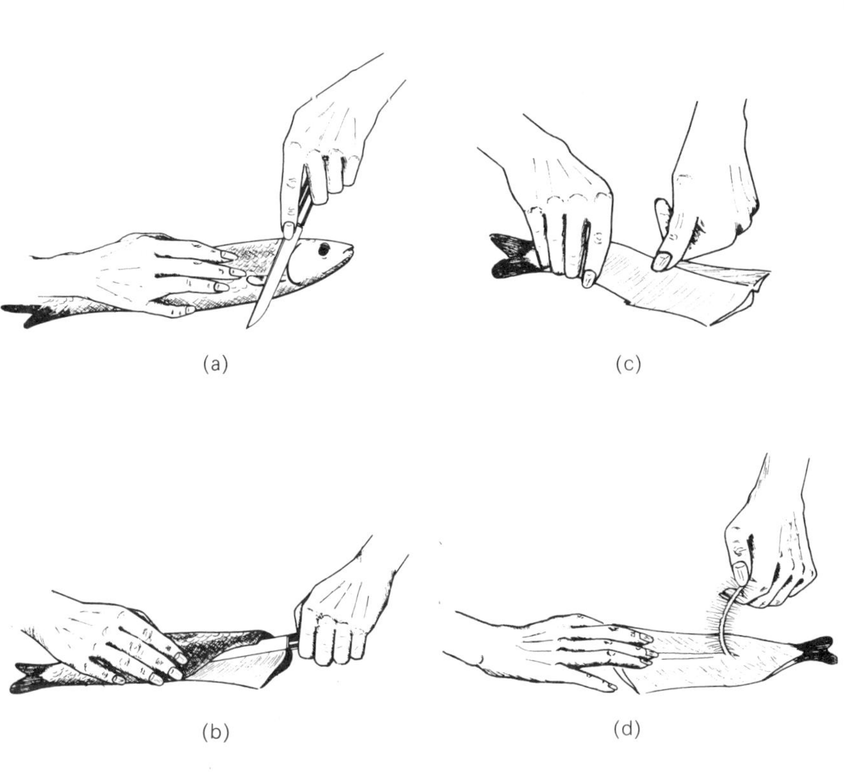

Fig. 18 Cleaning and boning a fish. (a) Cutting off the head. (b) Splitting the fish along its belly. (c) Running fingers along the backbone. (d) Removing the backbone and ribs.

Method 2

(a) Place the fish, either skinned or unskinned, flat on a wooden or Formica surface.

(b) Make an incision with a sharp pointed knife along the whole length of the backbone from head to tail.

(c) Next, insert the knife in this slit and carefully separate the flesh from the bone along its length, keeping the knife pressed lightly against the bone.

(d) Remove the fillets.

Quantity of fish to serve

This is given as a rough guide. In the first place it can be assumed that after preparing and cooking, the fish will be reduced to about half its live weight.

When serving fresh-water fish whole allow one medium-sized trout weighing about 1 lb (454 g) for each person. With filleted fish, an allowance of 6 oz (170 g) or a little more should be provided per person.

General points about cooking fresh-water fish

In general many types of fresh-water fish lack flavour, and in some cases they need to be served with plenty of flavouring and vegetables and herbs, such as onions, parsley and bouquet garni, and with piquant sauces. Straight boiling is the least desirable way of cooking.

In addition, some fish have a taste which, if not treated, is disagreeable to many people. Although carp, for example, are the subject of large-scale cultivation for culinary use in many countries and enjoy a reputation of being delicious, some people regard them as very poor food. This is largely because it has a marked muddy tang, probably because these fish grovel in the mud for their food and actually suck into their systems appreciable quantities of mud. Some people consider that tench are similarly tainted. In fact, when this fish is cooked it is recommended that the gills should be removed because they always contain mud.

Carp must not be cooked as soon as they are taken from the pond. They should be placed alive in a container through which running water is passing. This, it is claimed, removes any taint. Some cooks recommend that the fish should be fed for a few days in this fresh water with food that is not likely to contaminate their flesh. The same treatment can be used on tench.

Because of the lack of flavour of fresh-water fish, most cooks do not boil them but prefer to poach them. In general terms this means that the fish is placed into cold liquid, brought to the boil and allowed to simmer for about seven minutes for each lb (454 g). The liquid can be fish stock, milk, wine, cider, water or court bouillon. The latter might be fish stock blended with white wine, bouquet garni and seasoning.

Rainbow trout is excused from any criticism that its flesh is in any way tainted. It is regarded everywhere as a delicacy. But if this fish has any suspicion of a muddy flavour it can be dealt with in the same manner as carp and tench.

Recipes

Baked stuffed rainbow trout

Clean, scale, gut and dry two rainbow trout. Fill with veal forcemeat and close by sewing up. Place them in a baking tin covered with 2 oz (57 g) of dripping and bake them in a moderate oven, 350–375°F (177 to 191°C) (Gas Mark 4), with frequent basting. Next mix 1 oz (28 g) of flour with a similar quantity of melted butter and fry the mixture. When the fish is cooked remove it and put it on a hot dish. Strain the liquor in which the trout has been cooked on to the flour and butter mixture. Bring this to the boil, stirring it all the time, and add one dessertspoonful of capers, one teaspoonful of lemon juice, half a teaspoonful of anchovy sauce. Season to taste and allow it to simmer for a minute or two. Finally, pour it over the fish and serve.

Rainbow trout á la meunière

Clean two medium-sized rainbow trout. Season them, dip them in flour and fry in shallow butter until golden brown. Remove and sprinkle with finely chopped parsley, capers and lemon juice. In the meantime melt some more butter in the pan and heat until it turns pale brown. Pour on to the fish and serve immediately.

Rainbow trout fontainebleau

Roll four cleaned rainbow trout in seasoned flour and fry slowly in plenty of butter until golden brown. Put into a warm oven. In the meantime, remove the seeds and skins from ½ lb (227 g) tomatoes, chop them up and cook to a pulp with one dessertspoonful of tomato purée and a clove of garlic. Strain. Add to the butter in the frying pan a tablespoonful of chopped onion, a pinch of saffron, ¼ pt (142 ml) cream and the tomato pulp. (If necessary more butter can be added.) Bring to the boil, stirring. Pour over the fish. Serve very hot.

Smoked rainbow trout pâté

Mix 8 oz (227 g) of smoked trout (see Chapter 13) with 2 oz (57 g) of butter, the juice of one lemon and a finely crushed clove of garlic. Thin to a soft consistency with single cream. Garnish with olives, gherkins and lemon.

Baked carp

Wash and dry some fillets of carp. Spread with fish paste. Place them in a little melted margarine or butter in an ovenproof dish. Top with a mixture of breadcrumbs, grated cheese and seasoning. Bake in a hot oven (450°F/232°C) (Gas Mark 8) for about half an hour or a little longer according to the thickness of the fillets. Serve with parsley or caper sauce.

Stewed carp

Wash a large carp in vinegar and water and then cut into thick slices. Fry two or three small sliced onions in 1½ oz (42 g) of butter. Add stock, mushrooms, bouquet garni, a pinch of grated nutmeg and pepper and salt. When warm add the carp and allow to simmer gently for 30–40 minutes. Remove the fish and keep hot. In the meantime knead one tablespoonful of flour and ½ oz (14 g) of butter into a smooth paste and add this to the ingredients in the pan. Simmer and stir to produce a smooth sauce. To serve, put the fish on a hot dish, strain the sauce over it and garnish with mushrooms and croûtons.

Tench can be cooked similarly.

Baked tench

As well as the normal gutting and cleaning, the gills of tench must be cut out because they are always muddy. Sprinkle the fish with lemon juice and put aside for 60 minutes. Melt 3 oz (84 g) of dripping in a baking dish. Put the fish in and baste well. Season and add two finely chopped shallots or small onions. Bake for 25–35 minutes, according to size, in a medium oven. In the meantime make ½ pint (284 ml) of white sauce and add to it some chopped gherkins, a tablespoonful lemon juice, pepper and salt. Pour this sauce over the tench when serving.

12 Deep-freezing Fish

It has already been pointed out that there is a close season every year for coarse fish, which include carp, tench and brown trout, and that there is a movement towards establishing a close season for rainbow trout. During such times it is illegal to catch fish, even if the waters in which they live are in private gardens. This, however, presents no problem to the garden fish farmer if the fish are caught during the permitted period, deep frozen and stored in the freezer. Fortunately the storage time during which fish can be kept in prime condition in a freezer is approximately the same as the close season.

The process of deep-freezing food retards the action of enzymes and bacteria. The secret of deep-freezing is bringing the temperature of the food down to 0°F (−18°C), or even lower, rapidly. This is because meat, vegetables, fruit and fish contain a certain amount of water which is contained within the cells. By freezing quickly the ice formed is made up of small crystals, whereas when this process is carried out slowly large crystals form and, because water expands on freezing, these rupture the walls of the cells and it becomes impossible to restore the food to its normal condition on thawing. On thawing, the small crystals melt and the texture of the fish or other food is left unchanged. The enzymes and bacteria cease to be dormant and the rate of deterioration becomes at least as great as it is with fresh food. It must, therefore, be used as soon as possible after defrosting.

Types of deep-freezer

There are principally two types—the upright cupboard type and the chest type, both of which are thermostatically controlled. It is important to be sure that any freezer has an alarm fitted to it to indicate when the temperature inside it is rising because of power failure or mechanical breakdown. This warning system is often in the form of a battery-operated buzzer. Check this mechanism at frequent intervals to be sure it is in working order. If the alarm goes off, keep the lid of the freezer firmly shut, which will keep frozen food quite safely for up to 24 hours. If the fault is not obvious, call in the service engineer immediately. You can take out an insurance policy against loss of food due to power failure, and also for burglary.

A freezer normally needs little attention. It should be serviced annually. Defrost it every six months to keep it at an efficient working level. Do this in accordance with the manufacturer's instructions.

Quality of food to be deep-frozen

To be sure of complete success, fish must be processed not later than 24 hours after being caught, and the sooner the better. An exception to this is smoked fish, with which a delay of two or three days after catching can be tolerated. Even so, quick action with the smoking and freezing processes ensures a better-quality product when it eventually comes to the table.

Packaging for freezing

Everything for the freezer must be contained in waterproof and vapour-proof packages. If this is not done, deterioration will be accelerated and the fish will dry out. Also when there are strong odours, such as with smoked fish, other food in the freezer is likely to become contaminated. Special bags for packaging foods can be obtained from most large stationers and stores. It is most important to use the right type for the purpose. Fish fillets should be separated by a layer of thin polythene before packing, to prevent them sticking together.

Real success in freezing any food lies in securely sealing the package. First remove as much air as possible, either by pressing the bag or by sucking it out, using a drinking straw inserted into the package through a small hole formed by folding together the opening of the bag.

There are several ways of sealing freeze bags. Undoubtedly the most efficient way is heat sealing. This can be done by placing a thin strip of brown paper over the open end of the bag and applying a warm iron to it. But it is much easier to buy a special sealing iron and treated paper with a built-in strip. Wire tags can be used to close the bags, but they can work loose. Similar plastic tags can also be used.

Preparation of fish for freezing

Whole fish can be frozen straight from the water, or gutted. Freezing whole is a most satisfactory way, but gutting when the fish is icy cold after thawing out is an unpleasant task. It is probably better to gut beforehand.

The fish should be wiped with a damp cloth, then wrapped in foil or freeze paper and placed in a freeze bag, which should be tightly sealed.

If the fish is filleted each fillet should be wrapped separately and several put in a freeze bag. Large fish, cut into steaks, can be handled similarly.

Storage time

Good-quality fresh fish is quite safe in a freezer for from four to six months. The longer period applies to white (lean) fish and the shorter to the more oily ones. It is worth noting that salted fish generally has a much shorter life in a freezer than unsalted. In some cases it is reduced by two-thirds. This is particularly the case with oily fish, because the salt accelerates the development of rancidity.

Thawing

By far the most satisfactory way of thawing frozen fish is to leave it in its original wrapper in a refrigerator. This, however, takes rather a long time—3–4 hours for 1 lb (454 g). Not quite so good, but quicker, is to thaw it out at room temperature, when 1 lb (454 g) will be ready for cooking in about an hour.

13 Smoking Fish

Smoking foods as a means of preserving them has been carried out for centuries. Among the most popular foods preserved in this way was bacon, but in coastal districts fresh fish was also smoked.

To smoke fish it must be exposed to wood smoke, usually from shavings or sawdust. The wood most commonly used is oak. Sometimes apple-wood sawdust is used, but this tends to be more difficult to obtain. The manufacturers of modern domestic smoking equipment sell pre-packed sawdust. Experts and connoisseurs consider the hardwoods produce the best results, and that, of these, oak creates the best colour and flavour. Softwoods, particularly pine and spruce, give the food a rather acrid taste and odour.

What happens to fish when it is smoked

The usual commercial practice is to soak the split fish or fillets for 10–20 minutes in strong brine, which may or not contain a dye, and then hang them in an atmosphere of warm wood or peat smoke. The nature of the product largely depends on the length of time the fish is exposed to the smoke. If it is short, the cure is light and the taste rather mild; if exposed for a long time, the cure is heavy and the taste much stronger. The end product is largely a matter of personal preference, but the keeping properties of more heavily smoked fish are greater than those of fish which has been lightly cured.

Surprisingly little is known about the chemicals present in wood smoke, but there is evidence that the active ingredients are phenols and aldehydes together with tar and certain other compounds. Phenol (carbolic acid) is well known as an antiseptic; it is, of course, also a poison, but there is no need for alarm, because the amount present in the flesh of smoked fish is so minute that it is not injurious to the eater, but is ample to preserve the fish. The aldehydes include formaldehyde, which for generations has been used for preserving specimens. Analyses have shown that 'finnan' haddocks that have been commercially smoked, contain between 100–200 parts per million of formaldehyde and kippers up to 1000 parts.

At the same time, when fish is smoked the warmth dries it. The combined effect of the smoking process is to form a surface skin and to coagulate a superficial layer of protein. The presence of the phenols and the aldehydes slows the rate at which the destructive bacteria work, and the life of the fish as a wholesome edible product is prolonged. If it is stored at normal room temperature it will certainly be quite sound for at least three days, though fresh fish by that time would be quite definitely inedible. In a domestic refrigerator the time would be longer, but properly wrapped and put into a freezer it would keep for months. Despite the minor changes that take place in the composition of the flesh during smoking, the nutritional value of smoked fish is the same as that of fresh, untreated fish.

Modern ways of smoking fish

Very few people nowadays live in the sort of houses that have chimneys suitable for smoking foods. Simple, convenient, compact equipment has been developed for those who want to smoke food. Alternatively you can make a simple device yourself. Mrs Beaton's cookery book states that fish can be smoked at home by taking an old hogshead and stopping all the crevices. (A rather more easily obtained vessel today would be a steel drum with one end removed. This would not need sealing up.) Near the bottom fix a cross-stick on which can be hung the fish to be smoked. Next, cut in the side, near the top a hole large enough to take an iron pan to carry the sawdust. Turn the cask or drum upside down after fixing the fish to the cross-member. Position the pan through the hole. Place in it a piece of red-hot iron and cover it with the sawdust, which kindles and smoulders to give the necessary smoke.

It is much easier, however, to buy fish-smoking boxes. These range widely in their complexity and capacity and are stocked by many large stores, especially those that supply fishing gear or specialist kitchen equipment. The fish smokers differ in the amount of fish that can be handled at one time. The larger and more expensive ones can carry out other processes as well as smoking. A garden fish farmer who wants to smoke some rainbow trout for storing in his freezer during the close season will probably find one of the smaller, simple smokers quite adequate for his purpose, because the whole process can be completed in a few minutes, and quite a number of fish can be dealt with in a very short time.

The most practical thing to do is to smoke the fish out-of-doors in the garden or on a patio or open balcony, but if you have a very efficient extractor hood over your cooker it might be possible to carry out the process indoors. Some of the smaller models are intended for fishermen to prepare fish for their lunch in the open as soon as their catch has been landed.

The Abu smoker

This smoker, which was designed in Sweden, is small, very compact and portable. The overall dimensions of the smoke box are 11½ ins (29 cm) long × 6 ins (15 cm) wide × 3¼ ins (8 cm) high. It consists of a smoker with lid, and a grid, drip tray, scraper, fuel container and flame shield. These items pack neatly into the smoker for easy transportation and

The Abu smoker.

storage. Also supplied with the smoker is a recipe booklet and a supply of smoke-dust.

When using the Abu smoker, gut the fish but leave the skin and scales on. Rub in an ample quantity of salt. Sprinkle a thin, level layer of smoke-dust over the bottom of the oven. Put the grid in position, lay the fish (skin upwards) on it, and close the lid. Fill the fuel container with methylated spirit (or a tablet of solid fuel can be used). This is put centrally under the oven, and lit. After 8–15 minutes the fire will have gone out and the fish will be ready. Remember that the smoker works at a very high temperature so do not place it under a sheet of wood or plastic. Also, when refilling the container make sure that the flame is out and that the burner is cold.

Without the smoke-dust, the Abu smoker can function as a grill. It can also be used for smoking sausage and meat.

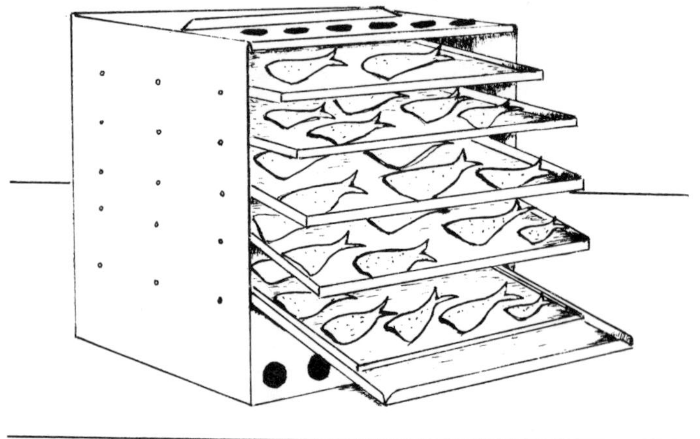

Fig. 19 The Homestead food dryer and fish smoker.

The Homestead food dryer and fish smoker

This more elaborate multi-purpose equipment can be used for blanching, sulphuring, drying and smoking. It consists of a metal cabinet with five trays, which slide out. It can be completely closed and the flow of air entering it can be controlled.

For smoking, the prepared fish is loaded on to the trays (see Fig. 19) and the air door closed. The apparatus is placed over an outdoor source of heat, such as a hot-plate or barbecue. The smoke-dust is placed in a shallow pan (or wrapped in aluminium foil punched with several holes), which is put over the heat source. The temperature is raised until the dust is smouldering without blazing.

Appendix: List of Commercial Fish Farms

The South and South-East

Aire Fishery,
3 New England Cottages,
Fobdown,
Alresford,
Hants.

Avon River Fisheries Ltd,
Bickton House,
Fordingbridge,
Hants.

Bossington Estate Trout Hatchery,
25 Bossington Estate,
Houghton,
Hants.

Broadlands Estate Trout Hatchery,
Broadlands,
Romsey, Hants.

Burford Trout Farm,
Burford Way Lodge,
Bedford Road,
Hitchin,
Herts.

Bury Hill Fisheries,
Lake View,
Old Bury Hill,
Westcott,
Nr. Dorking,
Surrey.

Froghall Aquatics,
Lambourne Road,
Chigwell Row,
Essex.

Hammer Trout Hatcheries,
Liphook,
Hants.

K. C. R. Howman,
Ashmere,
Felix Lane,
Shepperton,
Middlesex.

Kimbridge Fish Farm,
Kimbridge,
Romsey,
Hants.

Leckford Estate Trout Hatchery,
20 Longstock,
Stockbridge,
Hants.

New England Trout Hatchery,
Keepers Cottage,
New England,
Alresford,
Hants.

Stambridge Trout Fisheries,
Great Stambridge,
Rochford,
Essex.

Central England

Berkshire Trout Farm,
Hungerford,
Berks.

Bibury Trout Farm,
Bibury,
Glos.

Donnington Fish Farm,
Waterhead Barn,
Condicote Road,
Nr. Stow-on-the-Wold,
Glos.

The Midland Fishery,
Nailsworth,
Glos.

South Wales Fishery,
Michaelchurch-on-Arrow,
Kington,
Hereford and Worcester.

The West Country

Castle Hill Trout Farm,
Castle Hill,
Withiel Florey,
Nr. Wheddon Cross,
Somerset.

Cerne Valley Fisheries Ltd,
Abbots Leate,
Nethercare,
Cerne Abbas,
Dorchester,
Dorset.

Exe Valley Fishery Ltd,
Dulverton,
Somerset.

Gears Mills Fisheries,
Cann,
Shaftesbury,
Dorset.

Green Mill Fish Farm,
Probus,
Tresillian,
Truro,
Cornwall.

A. E. Hill,
Hooke Springs Farm,
The Millhouse,
Hooke,
Dorset.

Kerswell Springs Fish Farm,
Watercress Farm,
Kerswell Springs,
Chudleigh,
Devon.

H. McKin,
Golden Springs Farm,
Waddock Cross,
Wool,
Wareham,
Dorset.

North Dorset Trout Farmers Ltd,
Bourton Lodge,
Bourton,
Gillingham,
Dorset.

Pudleigh Fish Farm,
Combe St. Nicholas,
Chard,
Somerset.

Red Lake Trout Farm,
Somerton,
Somerset.

Road Water Fishery,
Watchet,
Somerset.

Rock Mill Trout Farm,
Rock Mill,
Membury,
Nr. Axminster,
Devon.

Samaaki Trout Farm,
Wolverton,
Zeals,
Warminster,
Wilts.

and at

Golden Cottage,
Bowerchalke,
Broadchalke,
Salisbury,
Wilts.

Sydenham Barton Fish Farm,
Lewdown,
Okehampton,
Devon.

Tracey Mill Trout Farm,
Tracey Mill,
Honiton,
Devon.

Trafalgar Fisheries,
Standlynch,
Downton,
Salisbury,
Wilts.

Wiltshire Trout Fisheries,
Winterbourne Houghton,
Blandford Forum,
Dorset.

East Anglia

Bayfield Fish Farm Ltd,
Glandford Mill,
Holt,
Norwich,
Norfolk.

Ken Smith,
Taswood Trout Fishery,
Mill Road,
Flordon,
Norwich,
Norfolk.

Southburgh Trout Farm,
Hingham,
Norfolk.

Thompson and Morgan
(Ipswich) Ltd,
London Road,
Ipswich,
Suffolk.

Water Mill Trout Farm,
Westgate Mill,
Louth,
Lincs.

Westacre Trout Fishery Co.
Ltd,
Westacre,
Kings Lynn,
Norfolk.

The North

Belleau Bridge Trout Farm,
Belleau,
Alford,
Lincs.

Dunsop Trout Farm,
Dunsop,
Nr. Clitheroe,
Lancs.

Ellerburn Fish Farm,
Ellerburn,
Thornton le Dale,
Pickering,
North Yorks.

Humberside Fisheries,
Cleaves Farm,
Skerne,
Driffield,
South Yorks.

Lake District Trout Farm,
Rydal Mount,
Gilcrux,
Cumbria.

Lakeland Rainbow Trout
Farm,
Mosser Hill,
Cumbria.

Lawsons Fish Farm,
Avondale,
Chapel St Leonards,
Lincs.

Lumb Mill Trout Farm,
Lumb Mill Farm,
Cross Hills,
Keighley,
West Yorks.

North Cheshire Fisheries,
45 Highfield Road,
Lymm,
Cheshire.

Poundsworth Mill Fish Farm,
Great Driffield,
Humberside.

Spion Kop Fisheries,
Warsop,
Mansfield,
Notts.

Staveley Lake Fisheries,
Copgrove,
Harrogate,
North Yorks.

Thimbelby Mill Trout Farm
(Horncastle Trout and Fish
Farm),
Horncastle,
Lincs.

Trent Fish Culture Co. Ltd.,
Mercaston,
Nr. Brailsford,
Derby.

Wansford Trout Farm,
Wansford,
Great Driffield,
Humberside.

Welham Park Fish Farms
Ltd,
Malton,
North Yorks.

Wales

Afon Wen Fisheries,
Afon Wen,
Mold,
Clwyd.

Brian G. Bateman,
Orielton Trout Farm,

Castle Martin,
Pembroke,
Dyfed.

and at

Vicars Mill Trout Farm,
Llandissilio,
Clynderwen,
Dyfed.

Chirk Fishery Co. Ltd.,
Chirk,
Wrexham,
Clwyd.

Draethen Trout Farm,
Draethen,
Glamorganshire.

St. Brides Hatchery,
Millbrook House,
St. Brides,
Netherwent,
Newport,
Gwent.

Whitebrook Fisheries,
Whitebrook,
Llanvaches,
Newport,
Gwent.

Wye Valley Fishery,
Llansantffraed House,
Bwlch,
Brecon,
Powys.

Scotland

Castle Fisheries,
Low Balantyre,
Inverary,
Argyll.

Game Fisheries,
Loch Fitty,
Dunfermline.

Howietown and Northern Fisheries,
Stirling.

Kenmore Fisheries,
New Galloway,
Kirkcudbrightshire.

Kincardine Fisheries,
Ardgay,
Ross-shire.

Loch Leven Fishery,
Kinross Estate Offices,
Kinross.

Solway Fisheries,
New Abbey,
Dumfries.

Thurso Fisheries,
Thurso East,
Thurso,
Caithness.

Index

Note : Page numbers in italics refer to illustrations.